DELAYED RESPONSE

JASON FARMAN

DELAYED RESPONSE

The Art of Waiting
from the
Ancient to the
Instant World

Yale

UNIVERSITY

PRESS

NEW HAVEN AND LONDON

Published with assistance from the Alfred P. Sloan Foundation.
Published with assistance from the Louis Stern Memorial Fund.

Yale University Press books may be purchased in quantity for educational, business, or
promotional use. For information, please e-mail sales.press@yale.edu (U.S. office) or
sales@yaleup.co.uk (U.K. office).

Set in New Aster type by IDS Infotech, Ltd.
Printed in the United States of America.

Library of Congress Control Number: 2018933273
ISBN 978-0-300-22567-9 (hardcover : alk. paper)

A catalogue record for this book is available from the British Library.

This paper meets the requirements of ANSI/NISO Z39.48-1992 (Permanence of Paper).

10 9 8 7 6 5 4 3 2 1

For Jonah and Noelle
In Memory of Al Pramschufer

CONTENTS

DELAYED RESPONSE

INTRODUCTION

Why would you send a text message without text? Five years ago, as I sat in the audience at a conference in Boston, an anthropologist of Japanese popular culture described an emerging practice among teens in Japan. A teen would send a blank text message to his or her romantic partner, and the partner's job was to respond with a blank message with as little time elapsed as possible. These blank texts would be sent throughout the day and established a rhythm for the relationship, as partners would respond quickly to each other, always being in contact without saying a word.

I was fascinated. The meaning of these messages was entirely dependent on how quickly someone responded, and that speed—or delay, if someone's partner wasn't paying adequate attention—had significance for their relationship. The content of these messages was *time*. These nudges showed that each person was present despite being physically distant from the other. Waiting for a response was an interpretive moment; it was something to which the Japanese teens gave meaning.

The existence of the blank texts later proved dubious, but the idea behind them led me to write this book and to explore moments throughout history where waiting for a message was central to the message itself. For me, the story of the blank texts moved into the realm of communication media lore, serving as a founding myth that became one of the most apt descriptions of our era's relationship to messages and time. Our expectations for a quick reply, and the ways that the time lag between sending and receiving messages is interpreted, belong to a long lineage of technologies that have kept people in touch. In each culture and in every era, people have sent messages and waited for a response. We do it now with our emails and text messages, and in the past people waited for messengers running or riding on horseback, letters sealed or stamped, telegrams transmitted across oceans, and news updates printed in the morning edition. The delay between sending and receiving a message is something people have always interpreted with anxiety, hope, fear, boredom, or longing. These interpretations are powerful tools for shaping the ways that we understand human connection and intimacy. These interpretations also help unlock innovation, as we speculate about the unknown and create new ways of exploring the universe.

How quickly and how consistently people respond to messages gives a pace to their social connection. When one of the partners deviates from this rhythm, the other person is left to interpret the silence. The gap in time requires the person to fill in the blanks with meaning—or, more accurately, to assign those blanks meaning. Here, silence is content. In the mobile media era, we may ask: Did the other person's phone run out of batteries? Was she called into a work meeting? Is he cheating on me?

As it turned out, no one I spoke with in Japan had ever heard of the blank text messages that teens there were purportedly sending. I traveled to Japan to interview college students who would have exchanged blank texts back in their high school days. When I brought up the idea behind the blank texts, I was greeted with blank stares. These students hadn't sent blank text messages in their lives. I sat down over lunch or drinks with my Japanese friends and colleagues, who found the story amusing and bewildering. Though they found the concept interesting and believable (Japanese teens were always coming up with inventive ways of using their phones), they had never heard of the practice. Despite speaking to dozens of college students, professors, and media professionals in Tokyo, I was never able to confirm that blank texting had ever been a part of Japanese mobile media culture.

After I returned from my trip to Japan, though, I was discussing my search for the blank text messages with colleagues at my university in the Washington, DC, area. Raffaele, my colleague who grew up in Italy, told me he participated in a similar kind of message exchange during his teenage years in Italy. He called it "ringing." Ringing, it turned out, was an incredibly popular practice among Italian teenagers in the years leading up to smartphones and social media apps. Throughout the 1990s and early 2000s, it was common for Italian high schoolers to have cell phones, but using them was costly: There were no "unlimited" text, voice, and data plans. It was too expensive to send a casual text: "I miss you." To get around the high cost of using a mobile phone, they used ringing, in which one person would call a boyfriend, girlfriend, or close friend and hang up after a single ring. The unanswered call was free and communicated a clear message. Raffaele noted that the most common sentiment a ring would convey is, "Hey, I'm thinking about you!" Seeing a friend's or a lover's name on the caller ID, it was the other person's job to call back and leave a single ring as quickly as possible.[1] There were no words, pictures, videos, or "likes" with these messages; instead, *time* was the message. These rings echo experiences shared by anyone who has used mobile media to find and maintain a love interest: When it comes to sending messages with the person we love, waiting itself bears meaning.

As high schoolers in Italy during the early 2000s, Raffaele and his romantic interest would keep in touch by sending each other these textless messages. Though they would spend nearly every weekend hour side by side at small movie houses, at the nearby park, and at each other's houses, they attended different schools and rarely saw each other during the school week. To feel connected during the time between classes, Raffaele would ring his love interest, who would send back a single ring almost immediately.

One Italian writer wrote that a ring could save your day or ruin it. Rings could connect us, "to make you realize that I was thinking of you, that I remembered your existence, that your existence was not indifferent to me and that I hoped that mine was not for you."[2] The ring conveyed all of this without a single word exchanged. The meaning of these blank exchanges was based on the time between messages and the history of a particular relationship. Once this version of blank messages entered into the anxiety-ridden landscape of teenage life, these rings created a mixture of deep emotional connection, frustration, and social expectation. Raffaele told me about times when he would exchange daily rings with his close friends, who, like his romantic interest, were people he would see regularly on the weekends but not during the school week. If there was a day when a friend didn't send or respond to a ring, Raffaele would be left to interpret that silence, feeling frustrated and hurt. Rings created a routine of keeping in touch for these teens, and they would ring each other every day, for years.

Ringing was a part of maintaining a relationship and confirming intimacy. Through the exchange of these "textless" messages, during a period of life when young people define their identities in part based on who their closest friends and significant others are, a return ring with little time elapsed verified that the feelings of closeness were mutual. As such, Raffaele felt this pressure of a quick response to his love interest each time a ring popped up on his phone. He was keenly aware that making his love interest wait for a reply communicated a lack of attention, so if Raffaele was occupied with some other task and unable to reply, the lack of an immediate response hung in the back of his mind like a weight. It was in these moments of delay that he noticed how burdened he felt by these exchanges. These moments of waiting afforded Raffaele the necessary pauses to reflect on their relationship and, ultimately, provided him with the opportunity to understand that their relationship had reached its conclusion. He ended the relationship shortly thereafter.

These rings and the blank text messages resemble some of the features that have been rolled out onto social media platforms, like the Facebook poke and the Grindr tap. These features—notifications such as "Javier has poked you" or "Mike tapped you 15 minutes ago"—have largely been met with confusion and annoyance by users, but are meant to allow people to communicate without text. The poke was one of Facebook's first features, affording an interaction without the need to write a message. Facebook, and its users who loved the poke, saw its power in its ability to let users give it meaning. The social media company is currently rolling out new features that may work in ways similar to the poke, allowing textless greetings with a single click. These are considered "frictionless" design and streamlined modes of communication. They require little effort to initiate communication, which can be their power or their largest flaw, as some users interpret pokes and taps as superficial attempts at connection.

No one could confirm that the young people in Japan were doing anything similar to these rings, pokes, or taps with their text messages. While the story of blank text messages may be apocryphal, it remains compelling, just as the Italian rings are fascinating. These textless messages confirm something that we know deep down: time is a medium that communicates. Though it may seem obvious that human communication must be analyzed by looking at the words, images, and videos we exchange, the very definition of "content" must also include time.

There is a long, rich history of the study of nonverbal communication that shows the significant impact it has on how we understand one another.[3] Some researchers have gone as far as to say that when humans communicate, 70 to 93 percent of our messages come through nonverbal forms. Studies of nonverbal communication focus on messages encoded in gesture, body position, clothing, facial expressions, and (yes) time. This focus on time as a mode of communication is known as chronemics. The anthropologist Edward T. Hall begins his book *The Silent Language* with an eye-opening statement about the role of chronemics: "Time talks. It speaks more plainly than words. The message it conveys comes through loud and clear. Because it is manipulated less consciously, it is subject to less distortion than the spoken language. It can shout truth where words lie."[4]

Scholars like Hall have argued that nonverbal communication never stops; while words or images may be sent or not sent, nonverbal cues are always being produced whether we want them to be or not. So even when we're not speaking, texting, or sharing images with one another, we're always communicating. What time *means,* on the other hand, is always contextual. It depends on the particular relationship and the situational context of that relationship. In the story of the blank messages or rings that the teens were sending, time similarly communicated but didn't always communicate the same things; instead, time is filtered through the many ways that people "read" one another.

Comedian and author Jessi Klein details a story in her life when the "pregnant pause" between sending and receiving a message communicated everything she needed to know. She had reluctantly entered a wedding gown boutique to purchase a dress for her upcoming ceremony. While she was excited about her wedding, she was less enthusiastic about buying a dress: "It's as if these dresses are designed to erase your individuality, leveling you into a universal symbol of femaleness, like that faceless woman wearing a triangle dress on the door of every ladies' restroom in America."[5] Under the pressures of friends and family, she ends up trying on over a hundred gowns. Ultimately, she makes an unconventional selection, which she says "looks like something a very slutty saloon owner from the Old West would wear." She takes a picture of it to text it to her friend, hoping to get immediate approval. "I don't get that," she recalls. "I get those three dots that you see on an iPhone when someone is texting you back. And they're starting and stopping a lot in that way . . . when someone has something deeply unpleasant to tell you and is really struggling."[6] Shrugging off this sign of disapproval, Klein texts the image to all of her friends to get their impressions. "Everyone is three-dotting it," she says. After months of dress hunting, she decides to ignore her friends' distaste and wear the dress anyway.

These kinds of pauses often communicate more clearly than words, and we give equal weight to these delays as we're getting to know people, especially in the world of dating and relationships. The search for love and sex in the mobile media age is rife with examples of how we read one another's use of time. Once people connect and send messages through dating apps or after exchanging contact information, the time lag between responses takes on significance. How quickly (or slowly) someone responds to a message is read as interest or indifference. After sending a witty message to a love interest, a person might see those same three dots that Klein observed, indicating that someone is writing back immediately, what has been called "the smartphone equivalent of the slow trip up to the top of a roller coaster."[7] When those dots disappear and no response is given, the time lag communicates volumes of information to the person waiting for the reply.

Feelings about these kinds of silences, delays, or three-dotted hesitations resonate with a wide array of people in the digital age who use mobile devices as a medium to find romance, sex, and love. These examples show both the powerful ways we can keep in touch by using mobile technologies and the kinds of challenges and disappointments that can emerge. Silences take on meaning, and we're often left to interpret them as nonverbal cues that communicate to us in sometimes vague and frustrating ways. Take, for example, the term "ghosting," used to describe an unexplained break between two people, usually in a romantic relationship, in a previously consistent flow of text messages. This interval leaves the person who was "ghosted" to interpret what prompted the "ghoster" to stop responding. The website Urban Dictionary has a fantastic definition of "ghoster": after meeting a guy, often online or through a mobile dating app, you believe you've found your soulmate "because he likes all the same music and all the same movies, he tells you how amazing you are and says he wants to have a serious relationship."[8] After sending a message, the rhythm might begin to slow down between messages. Then there is silence. Without warning, he stops responding to messages. "Essentially he is so sweet you cannot understand why he would not be answering unless something terrible happened like he died and you have no way of knowing. He becomes a ghost. He haunts your thoughts and you can never figure out what happened." Did the ghosted say something that offended the ghoster? Was the ghosted too pushy or too needy? Did the ghoster lose interest? Was the ghoster bored by the conversation? These questions are never answered.

Many of those I interviewed and surveyed while researching this book expressed the emotional toll of not receiving a response. In romantic relationships, silence communicated in very distinct ways, usually leaving those waiting for a response to feel hurt or slighted. Since a mobile device is personal—a small technology that sits in our pockets against our bodies, or constantly in hand as we move about, or a short reach away in a bag or purse—then we assume that a person will have seen our message at some point in the day. The students I interviewed and surveyed expected a response within around five minutes and started to get concerned after an hour passed. Part of this logic is based on our own daily practices with these devices, as we pull them out every time we are forced to wait or get bored. A recent study of mobile phone users found that people under thirty-four years of age check their phones an average of 150 times a day.[9] Another study showed that across the age spectrum, users click, tap, or swipe their phones an average of more than twenty-six hundred times in a single day.[10] This constant connection with mobile devices means that as a message is sent it connects with the recipient almost instantly and is probably read within six or seven minutes. Mobile users look at their devices so often throughout the day that they expect their partners or love interests to do the same. So when a quick response isn't received, users give meaning to this silence and feel ignored or hurt.

Yet the promise of communication technologies is that they will connect people at an ever-accelerating pace until the distance between us is completely bridged. Contrary to the feelings of anxiety people have while waiting for messages, most of the contemporary rhetoric around the digital age seems to argue that digital media users have arrived at the promised era of instant connection. As senior editor at *Wired* magazine Michael Calore argues, "Without the internet, we wouldn't expect instant gratification as often as we do. Not just the ability to get online answers immediately, or same day delivery. Because of the internet, the anticipation of waiting for things is largely gone."[11] This applies not only to shopping or to accessing information, but also to how we communicate. Globally, there are more than seven and a half billion mobile phone subscriptions, allowing the vast majority of the world's population to be reached, and to reach out, at all times. Since 2009, mobile users around the world have been using these devices more for texting and emailing than for voice calls.[12] While this shift allows people to stay in touch with a romantic interest all day long, weaving conversations into the in-between times of a day, it also puts waiting at the center of their social lives. Mobile users have chosen a medium—the text message—that has waiting designed into its very fabric. Though the mythologies of the digital age continue to argue that we are eliminating waiting from daily life, we are actually putting it right at the center of how we connect with one another.

From a certain perspective, the ways that we cultivate romantic relationships in the mobile media age have no precedent in previous generations. The expectations for how long a response should take have undoubtedly shifted in massive and important ways; however, I intend to trace how people have given meaning to the wait times for messages back throughout human history. New technologies speed up the process of sending and receiving messages. This acceleration shifts the overall experience of time in each era and human intimacy is often defined based on the ways that we are able to keep in touch.

In 1847, the Post Office Department of the United States made a change that would transform how people across the then-expanding country would interact: postage stamps.[13] Up until this time, sending a letter was a bit like making a collect call; if someone sent a letter from Georgia to New York City, the cost would be calculated and paid by the New Yorker. The uncertainty and expense were burdens on the recipient, and so writing letters was rare, warranted mostly to announce major events like weddings and funerals. Once the price of postage dropped to a standardized system of five-cent and ten-cent stamps, many people could afford to write and send letters. That meant that sending messages could happen more frequently, which in turn meant letter writing became a two-way communication. Now writers expected a response. One result was a nation that felt more connected, as people began to move west toward areas of the country that were dotted with emerging farms, tracks made by caravans moving to the West Coast, and the increasingly familiar sight of the local post office.[14]

Just after the U.S. Revolutionary War, in the late 1700s, letters sent within the United States would typically take forty days to reach their destinations. A few decades later, in December of 1814, a peace treaty was signed in Ghent, the Netherlands, that ended the War of 1812. Yet the British and American soldiers fighting the war—even those in command—didn't hear about it for weeks. The Battle of New Orleans was fought two weeks after the end of the war. Reports of this battle didn't reach New York for twenty-seven days. Five days after the Battle of New Orleans was reported in New York, word of the peace treaty arrived in the United States, forty-nine days after it was signed.[15] Due to the slow crawl of this news of peace, hundreds of soldiers died in a battle that had no impact on the war's outcome.

By the time of the Civil War, in the 1860s, soldiers were using the recently introduced stamps to keep in touch with family in astonishing volume and at equally astounding speed. Each month during the Civil War, the Post Office Department processed eight million letters. Some units were sending out as many as six hundred letters per day. And by this time, the Post Office Department had found ways to make processing and travel times shorter. Trains transported much of the mail and workers sorted the letters in the trains en route. This meant that many of the letters written during this period arrived in less than half the time of those sent during the Revolutionary War. Many even arrived within ten days. As this shift took place, people began to have different expectations about response times. Reciprocity was expected at a particular pace that matched the speed of the technologies for delivery. When such expectations weren't achieved, and were met with silence instead, people used those moments of waiting for speculative meaning making.

During the Civil War, Lottie Putnam was left behind in Nashville, Tennessee, as her boyfriend, Richard H. Adams, Jr., went off to fight for the Confederacy. Putnam and Adams wrote to each other regularly while he was away. Suddenly, in early fall of 1863 there was silence from Adams. Putnam continued to write him letters but received no response. In a letter of March 1864, she wrote, "Why have you been so remise in writing to me, have I offended you in any way? If so I'm innocent and would like to be informed of all injustice done to my 'little friend.' I beg you, implore you, beseech you, entreat you to tell me why you have not written, and why you have obliterated me from your memory." While slow responses in eras past would not have necessarily led the ghosted person to ask, "Why have you obliterated me from your memory," as messaging technologies and media allowed for more people to write to each other and receive quick responses, the expectations shifted. The next month, Putnam heard that Adams had been captured and was in a prisoner of war camp. They were able to exchange letters while he was in prison, and eventually they reunited to marry after the war.[16]

Once air mail came to the United States, late in World War I, a letter could arrive at its destination in only a handful of days. First-class mail could arrive in the addressee's mailbox within three days. Local urban delivery could be even faster. From the late 1800s through the mid-1900s, pneumatic tube mailing systems moved mail at speeds of thirty miles per hour across cities like New York, Philadelphia, Chicago, Boston, and St. Louis. First-class letters sent through the tubes arrived within an hour. Two correspondents could exchange more than a dozen messages in a single day.

As far back as the Louisiana Purchase, getting a message to its destination quickly had been of key importance to the country. After gold was struck in Northern California, keeping prospectors connected with their families in Pittsburgh or Baltimore was one way to preserve intimacy as the nation took bicoastal shape.[17] It was how the country could become a "united" states: Its citizens were connected socially, and messages were the key to that social cohesion.

The impulse to stay intimately connected with distant friends or lovers resonates with a society that uses mobile technology largely for sending messages. In our own era, which is similarly dominated by the metaphor of movement that characterized the western expansion of the United States—as people across the country move in and out of urban centers, commute each morning and evening, change jobs, move away from the nuclear family, or face the inability to own a home and live a stationary life—sending messages is how we cohere. It's how we maintain relationships despite the distance. Increasingly, messages come to shape even the relationships of those who are also near us. These are tools to keep in touch throughout the day. And these messages are being delivered at faster and faster rates.

We hate waiting for a response to our messages. In fact, we hate waiting in all its forms. For most, it signals a loss of time that could be used in productive ways. It often creates anxiety about how long we'll be left waiting, especially if we don't receive much feedback. Hospital waiting rooms or buffering icons are designed for waiting, but neither gives us a reliable indication of when that waiting will cease. In circumstances where one person makes another wait either by being late or by forcing him to "cool his heels" in the reception area, it can be seen as a sign of disrespect and affects the power dynamics in a relationship. Waiting can make us feel powerless and frustrated.

There's something perhaps deeper, even existential, about our disdain for waiting. As Harold Schweizer writes in *On Waiting*, our hatred of waiting might be linked to the deep fear that waiting is all there is.[18] Life, we fear, is not waiting *for something*—to find the love of our life, for a better career opportunity to open up, for the plane to begin boarding so we can finally get away on vacation—but might only be the experience of time passing. Our hatred of waiting may be a reaction to the existential crisis that all we might have in life is to watch time pass without our ability to do anything about it.

This sentiment resonated in the last words of David Cassidy, the incredibly successful platinum-selling musician and actor: "So much wasted time." Cassidy's words were particularly surprising because we tend to believe that success in life has a direct link with using our time wisely. Since he appeared to be a success, we assume that he probably didn't waste much time in his life. We often look down on time wasting as unproductive, linked to laziness or unfulfilled potential. We valorize productivity and people who can manage time well, be punctual, and make the most of every minute of their lives. In this vein, we try to avoid moments when we know our time will be wasted. Sitting in the waiting room at the Department of Motor Vehicles waiting for our number to be called or sitting all day in the jury pool at the courthouse not knowing whether we will be chosen or sent home . . . these stretches of time are maddening when we can't leave and all we can do is wait.

However, these feelings have also found their way into smaller moments of waiting, such as standing in line or waiting for a video to load on our browsers. I studied these attitudes toward waiting and boredom with my undergraduate students and asked them to spend one week of the semester charting—down to the minute—what they did from the moment they woke up until they fell asleep. What each student in this class of thirty-five had in common was that there wasn't a minute of the day when they weren't occupying themselves with *something*. These students didn't have down time. Even standing in line meant taking out their phones to respond to text messages or check email or play games. There wasn't a moment in the day when they were unoccupied. If boredom peeked around the corner, these students ran the other way by keeping themselves busy; and if not "busy" with some kind of task or reading the latest updates on social media or news sites, they occupied their time by playing a game on their mobile devices.

Digital media companies have picked up on these avoidance techniques, because our refusal to wait directly affects their revenue stream. For some companies, our refusal to wait leads us straight to their product, as we check our phones while sitting alone at the bar, flipping through apps that might occupy our time while we wait. For other companies, there is a loss of profits if customers are made to wait for their product. Amazon carried out a study to see what the impact of waiting was on their revenue. It found that, on average, for every tenth of a second that its customers are made to wait while using its website, Amazon would lose 1 percent revenue.[19] In 2006, based on customer demand, Google experimented with presenting more search results on each page, testing a page of thirty results instead of its standard ten. Suddenly there was a 20 percent drop in traffic among the group being tested. Marissa Mayer, then Google's vice president of search products and user experience, found that the tipping point was the delay in producing these results. The page with ten search results took 0.4 seconds to load, while the page with thirty results took 0.9 seconds. The half-second delay, a difference barely noticeable (if at all)—at least on the conscious level—was enough to dramatically affect site traffic.[20] Thus these companies have devoted enormous resources to speeding up their sites in ways that are often beyond a customer's awareness but have a direct influence on that customer's attitude and willingness to stick around.

In some instances, like the half-second delay of Google's search results, the wait time is not something we are keenly aware of, but it still affects how we feel about a product or a situation. Wait times are a bit like broken moments that call attention to themselves to varying degrees. A half-second delay doesn't hinder a person from using Google, but it's enough to make us feel that something about the technology is not quite meeting expectations. And once waiting becomes "slow, thick, opaque, unlike the transparent, inconspicuous time in which we accomplish our tasks and meet our appointments," as Schweizer writes, time has broken down.[21] Time, in this regard, is a bit like a tool that works so well that it incorporates itself right into your body and life. You no longer notice it until it breaks.[22] Wait times are the moments of breakdown when we become aware of time in distinct ways. We may notice it only in nearly imperceptible ways, like the half-second delay in a Google search, or it may gnaw on us in agonizing ways. We may feel like the philosopher Henri Bergson, who wrote, "It is we who are passing when we say time passes."[23]

Waiting, ultimately, is defined by the very fact that we notice it. It pulls us into an experience of time in unique ways. It makes us notice duration. For most people, this awareness of time is coupled with discomfort and annoyance.[24] Others, though, have seized on this slowed-down, attentive time as an avenue to a more deliberate life. Slow movements have emerged across spheres of life like food, scholarship, travel, fashion, and even religion. Many of these movements have attempted to reclaim these annoying wait times and instead insert them as an antidote for our sped-up lives. Yet many of these slow movements don't eliminate the ways in which waiting is either a privilege for the elite or a limitation forced upon the seemingly powerless. Waiting, rather than liberating some from the pace of the digital age, functions to remind them of the power they lack. Someone else controls their time, so they must continue to sit in the waiting room until called upon. These kinds of critiques of slow movements are at the center of scholar Sarah Sharma's research. In her book *In the Meantime*, she traces the ways workers' time is calibrated to the time of those in positions of power in a society.[25] A key question for Sharma's work is: Who is allowed to be slow and who isn't? She describes the scenes of slow-food gatherings, where wealthy people converge around plates of locally sourced food prepared at an unhurried pace. In the background are laborers (often people of color) hustling to wash plates, keep the kitchen clean, and take out the trash. For some of these workers, this is a second or third job. In many cases, the slow movement slows down for those who already have the resources to do so. For everyone else, life is lived at a breakneck pace except when someone makes you wait because his or her time is more valuable than yours.

The slow movement and the tech industry have something in common: Both are trying to reconfigure what it means to wait. Both want to use these moments of waiting to pull you in and shift your attention. Social media companies want you to spend those annoying minutes of waiting in focusing on your friends' posts that are strategically positioned next to advertisements. The "slow church" wants to counter the speed of the digital age by taking moments of pause as an opportunity for fellowship. These movements are about shifting the focus of how time is used. Sharma's provocative question asks *to whose time are we asked to calibrate?* That is, our sped-up lives were in sync with the technologies that have made instant, fast-paced culture possible. As we slow down, with what or whom are we supposed to be in sync?

Waiting here reveals its contradiction. It is both an opportunity and the thing that needs to be eliminated. For the tech industry, waiting and the attendant awareness of time passing provide the key moment to engage with their products. Digital media companies use waiting to their advantage in examples ranging from time-passing mobile games to announcements of the upcoming release of a new device that builds anticipation for it to hit the market. Conversely, waiting is the thing that has to be eliminated; it is the bug in the system. The mythologies of the digital age center around the idea that waiting is keeping you from obtaining what you want and holding you back from living a more fulfilling and productive life.[26] What emerges out of this contradiction is a particular orientation toward waiting that positions it as alternately a hurdle to our desires and an opportunity to use our time better. These attitudes ultimately support the current system to keep us busy to the point where we can't breathe. Our time is calibrated to a notion of efficiency that, in a single gesture, both demonizes waiting and preys on it as the opportune moment to occupy our attention.

We feel busy in the digital age. Life is hectic and we don't seem to have time to do the things we want to do. Waiting, in this context, is a waste of time. And time is our scarcest resource. Did people always feel this way?

As I began thinking about the ways that the Japanese blank text messages represented (even if mythical) something that had been true throughout history—that waiting is time we give meaning to, that waiting itself is a medium that takes on meaning—I wondered how the pace of life in eras past was affected by their technologies. Did the pace at which a message was delivered change the human experience of time? Did life always feel like it was accelerating because of the advancements in how people kept in contact with one another?

The tempos of life and the ways durations like waiting are experienced are always linked with the technologies of an era that regiment and divide time. From clocks to time zones, from trains to telegraphs, from pocket watches to mobile phones, lived time is directly linked to the technologies that shape and communicate that time.

To investigate this meaning, I move from the mobile age of text messaging back through time to explore the ways communication technologies have shaped the experience of time. To answer the question about how technologies have affected peoples' sense of time, intimacy, and their place in the world, I've gone on a kind of archeological dig; I've sought out unexplored corners of communication history to look at how technologies of the past changed notions of time and human connection.

In the journey to answer this question, I traveled to Tokyo to study contemporary text message habits of college students in this city. I then spent time digging through archives across the United States to study the pneumatic tube mailing system, a technology that allowed people to exchange messages throughout the day. I re-created maps of this system as it once existed in New York and spent time trying to track down its remains in order to understand the fascination this and other cities had with early versions of "instant" messaging. This system created an experience of rapid connectivity that mirrors our own feelings of acceleration and speed of communication.

I interviewed many experts, ranging from interface designers, who spoke to me about such designs of waiting as buffering icons, to astrophysicists at NASA and the Johns Hopkins Applied Physics Lab, who talked with me about the time lag for communicating with spacecraft at the edge of our solar system. I journeyed to the Jet Propulsion Laboratory in Pasadena, California, to see the Deep Space Network of massive satellite dishes that send and receive messages from newly explored regions of our universe.

I traveled to back roads of Virginia to visit the site where a Civil War soldier wrote his last letters to his wife back in Michigan. At this spot in Fredericksburg, the Union Army was badly defeated due to delayed messages that stalled the arrival of bridge supplies to cross the river. I traveled to London to visit the archives of royal seals that were attached to documents as a way for people to prove their identities across the distances a message would travel. I held seals and documents from the medieval period, more than nine hundred years old, that would communicate the identity and authority of a person in power, letting the recipient know that a reply should not be delayed. Finally, I spent time in Melbourne, Australia, with the Wurundjeri tribe in order to learn about the first messages humans ever sent through Aboriginal message sticks.

The range of examples I explore in this book demonstrates a key point: Waiting isn't an in-between time. Instead, this often-hated and underappreciated time has been a silent force that has shaped our social interactions. Waiting isn't a hurdle keeping us from intimacy and from living our lives to our fullest. Instead, waiting is essential to how we connect as humans through the messages we send. Waiting shapes our social lives in many ways, and waiting is something that can benefit us. Waiting can be fruitful. If we lose it, we will lose the ways that waiting shapes vital elements of our lives like social intimacy, the production of knowledge, and the creative practices that depend on the gaps formed by waiting.

I hope by the end of this book, you'll see waiting not as a burden, but as an important feature of human connection, intimacy, and learning. There is so much we can learn from waiting, if we only take the time.

1 WAITING FOR WORD

On my first trip to Tokyo, I came across a statue of a dog named Hachiko. Hachiko is legendary in Japan and has become a mascot of the city. The citizens of Tokyo see Hachiko everywhere. There are cartoons of Hachiko on the sides of vending machines. There are Hachiko T-shirts, stuffed animals, and tote bags in stores. Hachiko is more than a part of the "cute culture" that suffuses Japan; he represents how people in the city relate to time, patience, and waiting.

During this trip I stayed in Shibuya, a bustling neighborhood that is plastered with lights and filled with sound. Here, the statue of this dog is a main gathering site just outside the Hachiko Exit to the Shibuya Metro station. As I stepped out the door of the station with my luggage, I saw a host of young people gathered around the statue, some taking pictures with it, others just waiting. One person walked up and put a sash over the dog. This statue has become a site for gathering, as college and high school students send text messages, "Meet me at Hachiko." As I pulled my suitcase past the tourists, students, and families, several of them were scrolling through messages, waiting for friends to arrive.

These were the "smart mobs" described by media and communication writer Howard Rheingold. Rheingold details a major transformation that he saw outside Shibuya station, at the "Shibuya scramble," a legendary and hectic street crossing immediately next to the Hachiko statue. On a spring day in 2000 at the Shibuya scramble, Rheingold noted that people crossing the street there were staring at their phones instead of talking into them. This signaled a major shift in how they were using technology: "The sight of this behavior, now commonplace in much of the world, triggered a sensation I had experienced few times before—the instant recognition that a technology is going to change my life in ways I can scarcely imagine."[1] He goes on to note that these shifts of mobile media use, especially focused on how people use messaging and the internet on mobile devices, ended up being a catalyzing force in bringing dictatorships down and transforming areas of life from dating to politics to corporate management styles.

The statue of Hachiko overlooks this constantly changing site in Tokyo. This iconic dog lived in the 1920s and accompanied his

Figure 1. The statue of Hachiko just outside of Shibuya Station in Tokyo. Hachiko has become a symbol of waiting in Tokyo and represents the ways that waiting is a form of loyalty, love, and perseverance. Image © 2016 Jason Farman.

master, Hidesaburō Ueno, each morning to Shibuya station. Here, Ueno boarded the train to his work in the Department of Agriculture at nearby Tokyo Imperial University. The dog then returned at the end of the day to greet Ueno at the exit of the station. Day in and day out, this was the pattern: Hachiko would walk with Ueno to the station, leave, and then return at the end of the workday, watching the passengers exit the station while looking for his master. Then one day, during a lecture, Ueno collapsed and died from a brain hemorrhage.

Hachiko, oblivious to what had happened, returned to the station at the end of the day to wait for Ueno. The story goes that he continued to do this for nearly a decade, day after day coming to the station to wait for his missing owner. Only the dog's own death in 1935 ended his wait.

A decade after Hachiko's death, two atomic bombs were dropped on Japan, and people around the country were left waiting for word from their loved ones in the aftermath. In the prelude to World War II, Hachiko symbolized traits that were touchstones for the Japanese imperialists: a pure bloodline, loyalty, strength that dominates weakness—all characteristics attributed to Hachiko's breed, the Akita.[2] After the war, Hachiko took on a different tenor. He embodied the wait for loved ones who would never return. He signified the distance of a growing and increasingly urbanized country that would separate families as they spread out for work. He represented a new kind of loyalty for a country that was rebuilding itself with the ghosts of those who had left. In this life, they would wait.

In 1999 the Japanese national postal service circulated a survey for images to be used on a forthcoming set of stamps that "best represented the Japanese experience in the twentieth century." Hachiko was chosen for one of these stamps. While Hachiko symbolized the Japanese experience in the twentieth century, his statue is at home in twenty-first-century Tokyo, surrounded by people waiting and staring at their mobile devices. I spoke with several people who were waiting at Hachiko, curious to learn whether this symbol of Tokyo resonated with them and whether they thought people in Tokyo "waited differently" from people in other cities because of the ideals represented by Hachiko. Some were tourists who had no knowledge of Hachiko's story; this was simply an easy and recognizable spot to meet their friends. A Japanese girl in a long white dress was sitting just behind Hachiko, scrolling through the messages on her phone. The friend she was waiting for arrived, and after they greeted each other, they sat down on the cement planter that surrounded the statue. They were waiting for other people to arrive for a concert they were attending that night in Shibuya. One member of the group had suggested gathering at this well-known meeting location. This group, in fact, had never met face-to-face; they knew one another from an online fan group for the band that was playing that night. I asked the girl in the white dress how she felt, in light of Hachiko's story, about waiting for people. She responded, "I don't mind waiting as long as there is feedback from the people I'm waiting for. If time has passed and there's no messages about why the person is late, my patience is short."

A young Japanese man in his early twenties was wandering around the Hachiko statue repeatedly glancing at his phone until his friend arrived and they embraced in a big hug. His friend was visiting from Germany, and they had met at this statue several times before, when his friend lived in Osaka as a student. "Hachiko is the symbol of Tokyo," the young Japanese man told me. "My parents grew up hearing the story of Hachiko, and it's something that many of our parents have handed down to us. What Hachiko symbolizes to us is kind of natural. He represents something that's deep within our culture." The young man went on to note that with contemporary tools for keeping in constant touch and coordinating time down to the second, the attitudes about waiting in Tokyo have shifted. "For the Japanese, punctuality is a basic skill. It's a virtue. If we are good friends, we can wait and be patient with one another. But if I were meeting him for the first time and he was late, I wouldn't feel good about it. My impression of him wouldn't be good." After our short conversation, he and his friend got in line behind tourists to take a selfie together in front of Hachiko.

From time to time as the people I interviewed waited for their friends or relatives, someone would receive a text message update detailing where the others were and how much longer it would be before they arrived. I wondered why these people weren't simply calling instead of texting. I've been obsessed with this question since I began teaching university students early in the 2000s. I was shocked by how many text messages my students would exchange in a single day and equally surprised at how the "phone" in "cell phone" was just a historical throwback that remained mostly unused on their devices. They rarely made phone calls on their mobile phones. Since then, I've been invested in trying to understand the allure of the text message and how this connects to the history of message exchange more broadly. There is an extraordinary power in the text message in contrast with the phone call, and this power accounts for the meteoric rise of texting and the significant drop-off of voice calls that began right around the time I started working with university students.

The first reason these people who are gathered around Hachiko are drawn to texting instead of calling, just like my university students, relates to the way these technologies fit into daily life. When we make a call on a mobile device to another person's device, we are no longer calling to a place; we're calling to a person.[3] When I was an undergraduate student and called my girlfriend on her family's landline number, I knew that I was reaching her at home (and I tried not to call during dinnertime). Now, when I call my wife or close friends on their mobile devices, I have little sense of what situation my call might be interrupting. Will my call arrive during a business meeting? Will my call interrupt an important conversation? Did my wife forget to turn off her phone's ringer and will my call's ring echo in the lecture hall as she's teaching? In Japan, it's culturally frowned upon to talk while on the Metro. Since most public transportation in Japan maintains this social expectation, answering a call on a mobile device while standing in a crowded subway car is seen as a violation of the social contract. Thus texting provides the opportunity to send a message and connect with someone in ways that fit within certain social norms while allowing the other person to respond when opportunity arises.

These messages are also incredibly powerful because texting is a medium that allows us to compress a lot of meaning. Texting is, in Marshall McLuhan's terms, a very "cool" medium, which requires us to fill in the blanks (as opposed to "hot" media, which are filled with information, allowing more passivity for the receiver).[4] We use texts and the time of exchanging them to project our perspectives onto the interaction. Though voice calling is a cool medium in comparison with something like video calling, texting is cooler still. In its blanks we shape our expectations of a relationship. In Tokyo, which is surrounded by hot media like video billboards that blast sound onto nearby intersections and screens on surfaces throughout the city, the cool medium of text messages offers a welcome shift. This shift is appealing because it gives these users a sense of agency in contributing to the meaning of an exchange.

Finally, by weaving themselves into our busy lives, text messages can give the sense of constant connection without constant vigilance to a single exchange. This is the pleasure of multitasking that has consumed our attention. For example, one student I interviewed at a coffee shop in the Harajuku neighborhood of Tokyo was, during our interview, engaged in six different conversations. Her phone was dinging notifications throughout our interview, and when she responded, she would get a message back in under a minute. Students like this one feel connected with their friends and romantic interests throughout the day, via these "asynchronous" exchanges—messages that happen outside of real time. Asynchronous messaging is pleasurable: It makes people feel connected to many others throughout the day. Pleasure also comes from the stimulation aroused in the pleasure centers of the brain by the device's notifications and with the challenge of multitasking that keeps these users' attention bouncing among conversations and platforms.[5]

These exchanges develop a pace and a rhythm for a person's day and for a person's relationships. Texting allows others to bring us in sync with them, and our responses confirm that rhythm. These rhythms are entirely dependent on replies; reciprocity is an acknowledgment of receipt and a way to establish a pace for message exchanges. Without someone reciprocating our message, we are left disconnected. When messages receive regular responses, these technologies give a rhythm to daily life and to our relationships.

How the people around the Hachiko statue gave meaning to their wait times was different based on the rhythms already established with the people who were about to arrive. In this sense, waiting fits with the philosophical and linguistic idea of "difference."[6] The idea of difference has to do with how the written word takes on meaning, with attention not just to the letters but to the spaces between those letters. A word can take shape only if there's enough white space between the letters for us to decipher them. If there were no space between letters, the written word would simply be a jumble of indecipherable dark lines. Meaning depends on the space between the letters and the space between the words. Similarly, difference also relates to the notion of deferral, or the need for us to wait until we get to the end of a sentence before meaning emerges.

Waiting plays a key role in these ideas. The space between letters and between words is like the pause between messages we send. Waiting is the blank space or the silence between letters or words needed for meaning making. But it's not simply the space between the black text to a white page that creates meaning, it's the shape of the letters that come both before the space and after. We create meaning by being able to differentiate between the shapes of the letters and to identify the words they form. This is similar to the context that shapes the ways that we give waiting meaning. What comes before a moment of waiting and immediately afterward form the meaning of waiting. It's contextual.

For example, when one person simply never responds (think of the "ghosting" described in the introduction, and Hachiko waiting for his master), a source of trauma for this kind of action is that it affects the meaning-making process, creating a never-ending delay to meaning. It's like a half-completed sentence with nothing but a blank page following it. These kinds of traumas also apply to other contexts, such as crises, emergencies, and moments of uncertainty when people only hear silence and wonder about the safety of their loved ones.

I came across many of these stories of silence and waiting while interviewing people in Japan following the March 2011 Tōhoku earthquake and tsunami. This 9.1 magnitude earthquake hit offshore

of the northern part of Japan and was the fourth-largest earthquake
ever recorded. Its force had severe consequences throughout the
country. It led to a massive tsunami, which hit the Fukushima area
and caused a level-7 meltdown at the Daiichi Nuclear Power Plant
complex. Years later, many who had been evacuated from
the area still were unable to move back home. Most relocated
permanently.

In cities across Japan, transportation systems came to a halt
as people were evacuated from subway lines in preparation of
massive aftershocks or crumbling infrastructure. In Tokyo, most
of the people I spoke with commuted into the city for work and
traveled on the subway about an hour to an hour and a half each
way. Now they were stranded and had no easy way to get home.
One man emerged from the subway station in a neighborhood he
hadn't seen before and began the long walk home, which would
take him all day. By some reports, more than five million people
were stranded in Tokyo when the quake hit.[7] Since many families
were not together in the immediate aftermath of the quake,
people tried to reach out on their mobile devices to make sure
their loved ones were safe.

Japan seemed to be in a better position than any other country
to face such a massive quake: Its communication system was designed to allow nearly every person in the country the tools, right
in their pockets, to reach out and be reached at any given time.
Though the lines for voice communication were down and people
were unable to make calls, smartphones were ubiquitous, and with
them the ability to send messages. But the 9.1-magnitude earthquake toppled notions of instant communication and connection.

The tension between having technologies for instant communication yet being forced to wait for a response could be seen
clearly following the Tōhoku earthquake. This prompted me, with
the help of a friend who was a professor in Tokyo, to set up a
meeting with a group of ten people, each of whom had been directly affected by the earthquake or by the tsunami. We sat at a
long table that took up the entire length of a tiny Italian restaurant
in Tokyo. Japanese beer was served alongside glasses of wine;
noodles were served alongside pasta dishes. Only a couple of the

people there had ever met before. Most were either professors of media studies or students of communication. They were eager to meet one another and hear the stories of the immediate aftermath of the quake and tsunami. As I began hearing their stories, through a translator (which required patience and waiting in itself, as we spoke late into the night), they would get up, swap chairs, and mingle together. Their stories ended up being the most poignant example of what I had come searching for in Japan.

Miho woke up in the middle of the night. It was 4 A.M. in São Paulo, Brazil, and as she turned on her mobile phone, the blue glow illuminated her face in the dark room. She was still a bit jet-lagged. She tried to force herself back to sleep, but she was in a friend's rented apartment, a strange room in a strange country far from her home in Japan. She was still feeling the effects from the previous nights and the revelries that continued in the wake of the famous Carnival she had come to see. She had turned on her phone to help with the jet lag, to scroll endlessly through a social media feed expecting to find nothing of interest, which would lull her back to sleep. Her thumb slid down the face of the phone, and her sleepy eyes would slightly close with each scroll. The light was too bright from the phone, blinding her for a moment, so she opened the display settings and set the brightness as low as it would go.

Miho opened her email first. An acquaintance from South Korea had sent a message with the subject line "Is everyone ok?" Miho's brow tightened as she opened it. "Is your family OK after the earthquake? I'm worried about them since they live right in Fukushima! Let me know!" I'm sure they are fine, Miho thought. Earthquakes are a way of life in Japan. Since her youth, growing up in Fukushima, there had been tens of thousands of earthquakes in Japan, though most of them were not noticeable at all. Near Fukushima, according to the Japan Meteorological Agency Earthquake Catalog, there had been about twenty-five significant localized quakes of 6.0 or more on the moment magnitude scale during Miho's life in Fukushima. She thought, "I'm sure this person is just being kind. If anything terrible had happened, I'm sure I would have received more messages than just this one." Her email inbox was otherwise empty. There were no text messages on her phone.

But then her social media feed began to shift. Instead of the boring pictures she had hoped for of Instagram-worthy lunch plates or reposts of mundane headlines, her Twitter timeline became filled with news of a massive earthquake near Fukushima and speculation that a tsunami would follow.

Her father worked in the nuclear power plant right on the coastline of the town. So far, there was no news about the plant being in any danger. She wanted to text him to make sure he was all right. Calculating the time difference was easy, since there was exactly twelve hours' difference between São Paulo and Fukushima; it would be just after 4:00 P.M. back home. Her father would be at work. Once he entered those doors, he was unreachable, since cellular coverage inside the plant was nonexistent. She called her mother. The call didn't go through. She sent text messages and emails, frantically pecking away at her mobile phone writing sentences that seemed to take too long to write. She just wanted to hear that everyone was fine. She sent the messages, her phone making a whooshing sound to confirm that each email and text message was on its way. Miho wished her friend in the next room would wake up so that she could tell him the news and share all her fears. Instead, she just sat in the guest bed, feeling as far away from home as anyone could ever feel. The messages probably landed on the other side of the Pacific Ocean only milliseconds later. "After all, we are able to connect to each other instantly these days, right?" Miho wondered. But instead of seeing an instant response, she could only wait for word from home. She waited for word that everyone was all right. She feared the worst as she sat in the glow of her phone in her friend's quiet apartment.

Ren was not like her friend Yuka. Yuka stared forward with calm assurance. Ren was in tears as she sat at the wood-laminate lunch table in the gym of her middle school in Tokyo. The earthquake seemed to last forever and hit right after the students were finished practicing for their graduation ceremony. "I had to leave my mobile phone at home," she told me. "A few years back, all middle schools in Tokyo banned mobile phones from school. You can't even bring them through the door." Tokyo epitomized cutting-edge technology, and smartphones were pervasive, even among very young children. By the time Ren reached middle school, four years after the release of the iPhone, internet-connected phones were the norm. The Board of Education responded to the predictable classroom disruption by banning all phones, so after the quake, Ren had no way to call her mother to make sure she was safe and on her way to pick her up.

"I sat for hours at one of the lunch benches," she said, "while friend after friend was taken home by their parents. As each friend's mom or dad came to pick them up, my fears grew and grew. I just wished I could have called my mom. There was no way for me to send a text or make any communication with my family. I just had to sit and wait."

By this time, her friend Yuka had made it home to be with her brother and father, but they had not heard from her mother. The earthquake had struck at 2:46 P.M. Though Tokyo sat 230 miles southwest of the quake's epicenter, the violent shaking from the quake could be felt for six straight minutes. It was now after dinnertime, and Yuka hadn't heard from her mother. True to form, Yuka wasn't worried. "My family always taught me that I had to be self-reliant in this world; essentially, you live alone and you shouldn't rely too much on other people." Yuka's mother had several relatives who died when they were young, which made a big impact on her. She didn't want her daughter to suffer the same shock if a loved one died, so she raised Yuka with a worldview stressing self-reliance. Yuka told me, "I knew my mom was self-reliant and she'd be all right. At worst, I was with my brother and father, so at least I knew they were safe." She would try not to read too much into the silence from her mother. "All the phone lines were down in Tokyo, and they shut down the Metro. I imagined that she was just stranded and slowly making her way home to us."

The lights flickered in the City Hall of Fujisawa, changing the industrial walls from their usual dull color to an even duller shade of gray. People from nearby houses and businesses had gathered outside with the city employees who had evacuated the building. Once those standing outside were cleared to enter again, they filed through the front doors one by one, in an order that symbolized Japan's strict social rules about waiting in line. Once inside, they gathered to get information from city officials about the state of emergency across the country. They stood close to one another, many wearing puffy jackets to protect from the cold bite of March.

Kazuo was a university doctoral student who was working part-time at City Hall when the quake hit. Dressed in a pressed blue shirt and thin-rimmed glasses, he grabbed a remote control from the table in the lobby and turned on the television. All heads seemed to snap in the same direction, looking to the screen for information. All they got was static. Older citizens shouted instructions at Kazuo to change the channel. He fumbled through a few stations unsuccessfully before graciously handing the remote over to one of the elders nearby. The man who was given the remote got frustrated after a minute of searching, shut off the television, and tossed the remote onto a nearby chair.

Kazuo's girlfriend (later his wife) worked in a high rise in Tokyo. "These high rises are something the city is quite proud of," a local friend told me. "Since earthquakes are so common here, the ability to build a safe building as tall as possible is seen as the pride of our engineering." Here, an hour south of Tokyo by train, Kazuo was responsible for the city's policies regarding digital life and family, focused especially on how families used mobile phones. Suddenly, as people were trying to communicate with their families in the immediate aftermath of the quake, his research started to look very real and immediate. People began to turn to him for help. The internet was being flooded with messages, phone lines were down, and the city's transportation system had ground to a halt. Families were trying desperately to connect, as schoolchildren like Ren were stranded in lunchrooms and parents were unable to call their children's schools to report that they were miles away with no means of transportation. Emails were sporadically getting through, but some recipients were unable to respond. People weren't sure whether their messages were being delivered, since most didn't come with the ubiquitous "Read" mark that now accompanies every text message sent over Japan's popular LINE app on smartphones.

Kazuo carried two phones, one in each pocket. One was an old-style phone called a *garakei*.[8] It was a black, sleek flip phone used mostly for simple tasks like making calls and sending texts. The other was a smartphone. With both phones, he sent text messages to his girlfriend, but he got no immediate response. With his *garakei*, he sent a message to his parents, who lived north of Tokyo and closer to the hardest-hit areas of the country. He also sent a text message to his brother, who worked as a manager and researcher for a pharmaceutical company in Fukushima. He sent this message last, moving unintentionally across the geography of Japan from those who lived closest to those farthest away. For Kazuo, this mirrored his relationships: While he spoke to his girlfriend throughout the day, he hardly talked to his brother. As he sent his message, he imagined his brother standing on the factory line that produced medicine during the quake, wondering how he fared.

Since the television was of little use to those in the crowd, they began to circle around Kazuo and his smartphone. Kazuo was lucky with his phone service; the mobile providers of most others at City Hall that day throttled their data to nearly zero in the wake of the earthquake. The companies did this to make sure their systems were not overwhelmed. But Kazuo used one of the least popular mobile providers in the country. Because his network was not packed with other customers competing for limited data, he was able to use his device freely. "Any news?" the old man prompted. Kazuo looked up at the crowd that had accumulated around him, unaware of its presence until the man had spoken. "Let me take a look," responded Kazuo. For the next few hours, as his phone's battery gradually drained, Kazuo was the default news source for those in City Hall. "What's happening in Sendai? My sister is a schoolteacher there," one young woman shouted over to him. With each request, he would scan the news sites and Twitter for any relevant information. "Can you email my parents for me?" another woman in her thirties asked, holding up her dead phone as proof of her need.

Then a notification popped up on Kazuo's screen that his girlfriend had just posted on Facebook. As he looked at her photographs of the damage to Tokyo—with a sigh of relief that she was safe—he got another notification of a text message from her almost immediately afterward. She was fine, busy working to help the people at her advertising agency. It had been three hours since he had sent his initial text to her; she usually sends a response in a matter of minutes. But since he had been occupied as the main news source for City Hall, he hadn't noticed the hours pass by as quickly as they did.

He turned again to the news, where a video was streaming of conditions near Fukushima. NHK news service, Japan's public broadcasting organization, showed video taken from a helicopter as the massive tsunami waves were coming ashore, a black wall of water that washed away houses on impact. A crowd of people gathered around the glow of Kazuo's mobile screen, watching the unstoppable wall of water catch escaping motorists in their cars, lifting vehicles into a swirling mess of broken wood, boats, roofs of houses, and mud. People gasped at the horror of the spectacle. Kazuo's breath grew short, as he thought of his brother. "Is no news good news?" he wondered. "I wish he would just message me back."

Miho's story about learning of the earthquake in a dark apartment in Brazil resonated with me because it captured many of the features I've heard countless times about waiting for messages. First, Miho was distant from her family. For Miho, long-distance communication with her family was a familiar practice, one used to maintain a kind of cohesion with them after she left Fukushima for Tokyo ten years before the quake hit. Messages became a way for daily life to be shared even if there wasn't a way to see each other face-to-face. While in Tokyo, she was able to establish a sense of connection across the distance between her new city of Tokyo and her family back in Fukushima. Yet in March 2011, being on the other side of the planet in Brazil presented new challenges to that sense of connection. The modes for communication that were typically used to keep that sense of intimacy between her and her family had been severed. Phone lines were down and messages weren't being responded to.

So, Miho got up and packed her suitcase. The friend she was visiting in São Paulo was now awake. He booted up his laptop to help Miho get in touch with people and stay up-to-date with the latest news coming out of the hardest-hit areas of Japan. The news organizations were not streaming any stories. Searching around other websites, they came across a young man's video feed from Hiroshima, which was so far from the epicenter that it was relatively unaffected by the earthquake. He was streaming a video feed of the news from his television set. "Usually, this kind of activity is banned," Miho told me. "But the video service, Ustream, along with NHK, decided to allow him to continue streaming their broadcast." Through the streaming video feed of the television, Miho and her friend both learned about the devastating tsunami that had hit the coastline of Fukushima, which only raised Miho's fears about her father. She heard about the evacuation of her neighborhood and speculated that her mother was at this evacuation site, unable to make a call or send a message. Miho's flight home was scheduled for later in the day, so she decided to get to the airport early and continue her attempts to get in touch with her family in Fukushima. She still had heard nothing from home as she boarded the first leg of her flight from São Paulo to New York. People on Twitter were furious that no news was being shared about the state of the nuclear plant. Everyone speculated as they waited for updates.

While Miho walked through the international terminal at JFK airport to her connecting flight, her phone rang. It was her mother. It had been hours, but Miho finally got word from home that her family was safe. Her father had taken a lunch break from work before the quake hit, but had to return to assess the damage after the earthquake. Her mother had indeed joined her neighbors at the evacuation site, but the nuclear plant had its own evacuation location. As a result, her mother and father were now separated and were being told by authorities that they needed to leave the area. Though each had a mobile phone, sporadic service and the inability to charge their devices disrupted their ability to stay in touch.

"I heard that the hospital near the coastal area of Fukushima is completely devastated," Miho said to her mother. "How will Dad get his dialysis? Doesn't he need to find a hospital in the next few days?" Her father's kidney problems required dialysis three times a week. Her mother said that they had agreed to meet at Miho's aunt's house in Ibaraki Prefecture, where her father would be able to get his treatments. What is typically a two-hour drive south from Fukushima ended up taking more than a day for her mother, who had to wait in long lines at gas stations to get a ration of gas for her car. The journey ended up taking her father four days. Her parents were separated for the entire trip out of Fukushima. In the coming weeks, they would move in with Miho's sister in Tokyo, and, finally, Miho would get to see her parents face-to-face. Her parents eventually moved north of Tokyo, and they have no plans to return to Fukushima.

All the people I spoke with in the restaurant in Tokyo had relatively good outcomes to their stories of the earthquake. Everyone made it home, and each family was safe. Ren's mother, who was stranded by the Tokyo subway system shutdown, eventually made her way to the middle school to pick up her daughter. She had been unable to call the school because her mobile provider had purposely reduced voice calls by 90 percent to make sure its system was not overwhelmed. Yuka's mother had faced a similar challenge, but eventually she made it home, too.

Kazuo's parents called him four hours after the earthquake, but at that point no one could reach his younger brother in Fukushima. This, however, wasn't unusual. His brother was known for late responses, so they didn't expect a quick reply. At 1:00 in the morning, Kazuo's phone alerted him to an email from his brother, telling him that he was all right. In my conversation with Kazuo, he told me, "Every few days, we'd get an email from my younger brother. But whenever my parents tried to email him or message him, he'd never reply. Whenever he wanted to talk, he'd reach out, but it never worked the other way around." Kazuo's brother stayed in the Fukushima area, just outside the region of the city evacuated for an extended period, working for the pharmaceutical company. The earthquake caused extensive damage to the part of the building where the factory line was located. Later that summer, Kazuo's brother returned home to Tokyo. When the brother arrived at his parents' house, they saw that their youngest son had lost an incredible amount of weight. "But he wasn't aware that he'd lost so much weight," Kazuo told me. "The others in his area had such a hard life, and his factory was deeply affected and the assembly line wasn't working well. So he had to work very hard and had a hard life after the earthquake. He didn't tell any of us how bad things actually were. We didn't realize how difficult life was for the people in the Fukushima area until he walked through the door and was hardly recognizable."

The silence from Kazuo's brother was interpreted in a specific way. Though silence was not unusual, Kazuo and his parents believed that the younger brother would reach out if he was in any sort of danger or difficulty. His lack of contact signaled to them that he was thriving despite the difficulties at the pharmaceutical company. "We always believed that no news was good news when it came to my brother," Kazuo said. The gaps between messages are significant based on the contexts that come before and after; that is, they *signify* something to the people who encounter those gaps in communication based on their relational and situational context.

These interpretations can often be wrong, as in the case of Kazuo's brother, who used silence as a strategy to not worry his family. His family interpreted this as if everything was fine. Not until he walked in the door were the real consequences of his situation revealed. Ultimately, Kazuo's family used its limited contextual clues to interpret the brother's silence: He was often silent, making this a norm that did not trigger fear in family members who had limited understanding of what has happening in his neighborhood and his workplace in Fukushima.

Kazuo's expectations were different for his brother than they were for his girlfriend. Time assumed different meanings as he waited to hear from each of them. Though he and his girlfriend communicated regularly, Kazuo did not have the expectation of a quick response from his brother. This expectation was complicated by the fact that his brother was in the hardest-hit region of the earthquake and tsunami. That context made the silence across the hours troubling, though he relied on their communication history as an explanation for not having heard from his younger brother. The multiple ways that time communicated were affected by the relational and contextual specifics of the moment. With no messages coming from his brother, Kazuo had to rely on the non-verbal communication that was available: time and waiting.

The earthquake disrupted the rhythms of communication for these relationships, and the silence imposed on them during these wait times shaped each person's sense of being connected with loved ones. For waiting to take on meaning as it does in our era, a certain level of synchronicity is required.[9] We are synchronized with the people in our lives through our ability to stay in touch with them. We are synchronized with these people and with the technologies that keep us connected. These technologies also give us a sense of synchronization with events happening around the world, connecting us to people in distant places. In crisis moments throughout history, word of critical events spread at a pace dependent on the communication media of the era. This is as true in the mobile media era as it has been for previous technologies for delivering messages.

For example, in the early years of the United States, news of George Washington's death on December 14, 1799, spread at a remarkably slow pace. He died just outside Washington, DC, and it took a week for news to reach New York City. It took nearly a month for the news to reach St. Louis. Now, as events happen that shape our national or global identities, we expect the information to come instantly. Our experience of "now" is distinct from other eras in this regard.[10]

This expectation of synchronicity is linked to the dominant communication technologies as well as modes of timekeeping of an era. In our digital age, many of us are on the same clock. Our mobile devices and digital technologies are all synchronized to the same atomic clock held at the U.S. Naval Observatory in Washington, DC. What this ultimately means is that each of us is not only synchronized down to the minute, but are actually synchronized down to the level of the nanosecond. With this time being available on the same device we use to communicate, we have expectations around how that time gets used in the ways we coordinate and keep in touch.

At the U.S. Naval Observatory, time is kept by several kinds of atomic clocks and then distributed to such utilities as the Global Positioning System (GPS) and the Network Time Protocol (NTP), which keeps time for online devices and browsers. According to Demetrios Matsakis, the chief scientist for time services at the Naval Observatory, if you put all of the clocks together at the Observatory and consider them a single clock, they constitute "the most accurate measuring device operationally ever created by mankind to measure anything."[11]

From these accurate measurements, time is then distributed to networks of clocks around the world. For GPS satellites, the time is dictated by the U.S. Air Force, which matches its time to the Naval Observatory's measurements. This time kept by GPS satellites is distributed to cell towers around the world, then handed off to every mobile device with a cellular signal. In essence, the timekeeping technology of the mobile phone is now accurate to the second in a way that no other device has ever been. More important, it is distributed to the majority of the world's population, making us all in sync down to the nanosecond. The excuse used by previous generations—"Sorry I'm late; my watch is running slow"—just doesn't fly in a world where everyone has the same exact time in a pocket or a purse.

Synchronization gave us a new orientation toward time and timeliness. There is a back-and-forth relationship between a technology that tells time accurately and the cultural need for accurate time. These two forces feed off each other and transform the ways we perceive time and waiting. As geographic distances began to be linked by modes of transportation and communication that sped up the exchange of messages and goods, there was an increasing need for synchronicity. As Alexis McCrossen writes in *Marking Modern Times*, which details the growth of ever more accurate technologies for timekeeping, "When it took a long time to traverse even small distances, the differences in time between places did not much matter. Each place had its own time, set by the sun and the stars."[12] This all changed as those distances were bridged. The technological forces that brought distant communities together, most notably the train, also brought about a need for those communities to be on the same clock. Being on the same time through accurate clocks and newly established time zones allowed people in these linked places to coordinate, collaborate, and create schedules for modes of transport between these locations. For McCrossen, this shift is what ushered in modernity. This shift carries over into our expectations today, as our synchronicity leads to the anticipation of instant connection with one another, especially in the wake of a natural disaster or crisis event.

Our own mobile devices have become the vehicles for our way staying in sync with the standardized time maintained at the U.S. Naval Observatory. They are descendants of the pocket watch in their mobility and the personalization of time. While we could manually synchronize pocket watches with clock towers, street clocks, or even time balls that would drop at noon in some cities, our mobile phones are all already synchronized, with no effort on our part. As a result, we have the most accurate measure of time ever created in our pockets and bags, and on our wrists, with smartwatches. This shift signals a distinct era for waiting and how time is regimented. It carries on the traditions of further segmenting time that affect everything from how a workday is spent to the body's rhythms of sleep. It extends those traditions by creating new expectations for timeliness and productivity. It also shifts the human experience of time; that is, our perception of time has shifted as each of us is synchronized with one another.

This kind of synchronization in its global reach connects us and also creates expectations for how we use our time with one another. It sets expectations for timely responses, since our devices are with us at all times. For the people I interviewed in Tokyo, the pervasiveness of the mobile phone in the culture symbolized constant synchronicity and connectivity. It meant that everyone in this technologically advanced society should have been able to reach out and send word instantly. Instead, they waited.

What it means to wait in an era of such synchronicity is dramatic. While contexts shift these meanings in important ways, and while each of us sets expectations for how quickly we as individuals will respond to certain people, the backdrop of all of these interactions is a new human sense of time that is globally distributed.

How have these rhythms and expectations changed over time? In what ways does our moment share a kinship with other moments of synchronicity, acceleration, and particular modes of waiting?[13] The synchronicity of our mobile media era, along with the accelerated rhythms of daily life, has made waiting seem like a bug in the system, but I have found many parallels to other communication technologies of the past. Here, these media that synchronize us at a particular pace also shift expectations around time, delay, and notions of instant connection.

2 INSTANT MESSAGES AND PNEUMATIC TUBES

In his short 1949 book *Here Is New York*, E. B. White wrote about the tangle of pipes, wires, passageways, and communication media under the streets of the city that allow daily life to happen. "It is a miracle that New York works at all," he wrote three years before publishing *Charlotte's Web*. "The subterranean system of telephone cables, power lines, steam pipes, gas mains and sewer pipes is reason enough to abandon the island to the gods and weevils. Every time an incision is made in the pavement, noisy surgeons expose ganglia that are tangled beyond belief." One of these ganglia at the time was the system of pneumatic tubes that shot mail around the city at thirty miles per hour between post offices. White wrote, "When a young man in Manhattan writes a letter to his girl in Brooklyn, the love message gets blown to her through a pneumatic tube—*pfft*—just like that."[1]

Half a century later, at the height of the dot com boom in the late 1990s, technology entrepreneur Randolph Stark was walking through his Wall Street neighborhood and saw crews digging up the street—the "surgeons" in White's description—to lay fiber optic cables between the banks and the stock exchange. Stark was on his way home from a tech gathering in Manhattan. Earlier that evening, over drinks, someone had mentioned to him that New York used to have miles of pneumatic tube lines that shot up to twenty thousand letters a minute between post offices. On the walk home from the meet up, as Stark was looking at the heavy machinery digging up the concrete and asphalt, it all seemed like an unnecessary amount of labor and cost—at about a thousand dollars a foot—when the infrastructure was already there just a few feet beneath the hole that was being dug anew. He envisioned

Figure 2. The ganglia of pneumatic tubes beneath the streets in New York City at the intersection of 17th Street and Sixth Avenue, 1906. Image courtesy of the National Archives, Washington, DC.

a new business venture: running fiber optic cables in the old pneumatic tubes that had been laid in the 1890s. This would allow companies, apartments, and office buildings to get faster internet connections, since the data would have less distance to travel. It looked promising, yet he had no idea who owned the unused tubes under the city streets or what their condition was. So, he went into the archives looking for answers.

As Stark sat among the gray archival boxes in the New York Public Library and the National Archives in Washington, DC, it was an odd juxtaposition to plan for the technology that would shape the future by digging through hundred-year-old documents about an obsolete mail system. However, there were many parallels between the fiber optic age of the internet and the era of the pneumatic tube mail system. Like our own moment, the age of pneumatic tubes created ways for people to send messages at unprecedented speeds. Starting in 1897, twenty-seven miles of tube were laid underneath the streets of New York City in order to use compressed air to shoot canisters of mail around the city between post offices. The pneumatic tube mail system, which pushed brass canisters that could hold six hundred letters each, were popular in Europe before their launch in the United States. Starting in the late nineteenth century, pneumatic tube systems became important to mail delivery in Philadelphia (the first city in the United States to start using pneumatic tubes), New York, Boston, St. Louis, and Chicago. The canisters—or "carriers," as they were called—would leave the post office every ten minutes and would fly underneath the city streets, able to move a message from Times Square to the General Post Office in three minutes even in a heavy snowstorm. With these tubes running underground and connecting people in new and technologically advanced ways, the era felt as if the possibility of instant connection was now at hand. Noticing the parallels between the pneumatic era and the fiber optic one, Stark asked himself, "Why don't we have pneumatic tubes running to every house? Why did that never happen?"

There have been moments in history when technologies allowed us to connect with one another at unprecedented speeds. These moments gave people the ability to send messages at rates

that seemed to eliminate waiting altogether. The rise of the pneumatic tube mail system was one such moment. In cities like New
York, the two lovers whom E. B. White described could exchange
more than a dozen messages throughout the day, heralding the
notion of "instant messaging." Cities across the country were
clamoring to install pneumatic tube mail systems. Sending mail
in canisters pushed by compressed air under the streets of a city
was seen as the essence of being cosmopolitan and modern. The
pneumatic tube system was not simply an efficient way to deliver
mail and packages across cities with congested streets now packed
with automobiles; instead, it was a symbol of modern life. Pneumatic tubes represented a technological leap forward.

This underground placement served two purposes. First, it
was practical, allowing the message canisters to be sent throughout
the city without interrupting life above the surface. The ability to
send messages without dealing with the crowded city streets or
severe weather was one of the main selling points of the pneumatic tube system. It could deliver consistent speeds regardless of
how congested or impassable the streets got above.

Second, by being out of view, the system allowed the imagination to create a mysticism that could be totally disconnected from
the physical reality of the pneumatic tubes. For example, a newspaper cartoon from 1915 that advocated for extending the system
shows a clunky mail car stuck at a bridge crossing while a missile-
shaped canister filled with mail shoots through a tube under the
river. The cartoon contrasts these with captions "What We Have,"
next to the mail car, and "What We *Ought to* Have" next to the
mail-missile. As communication scholar Esther Milne has argued,
the invisibility of the inner workings of these technologies supports
a myth that communication doesn't need a medium. Messages,
the idea goes, are about connecting mind to mind rather than
being subject to the limitations of a particular medium or series
of pipes that may inhibit or alter the ways that we communicate.[2]
So, out of view, these hidden technologies allow us to engage in
the fantasy that wherever we are, we can seamlessly connect with
another person and share our deepest thoughts. The words, not
the medium or the time it takes to send and receive a message,

Figure 3. A drawing published in the *Bronx Home News* on December 5, 1915, comparing the clunky automobile and its limitations with the futuristic (and militaristic) missile of the pneumatic tube canister being shot under the Harlem River. Sending mail in canisters pushed by compressed air under the streets of a city was seen as the essence of the cosmopolitan and modern. Image courtesy of the National Archives, Washington, DC.

are the content in this scenario. If people are able to simply share their thoughts, the medium should not affect that message.

In the era of the pneumatic tubes, "instant connection" through a messaging technology had a strong cultural allure, regardless of whether the systems could actually connect people instantly and seamlessly. Instantaneous communication is still an enormously powerful concept in our culture. It's the motivator behind that feeling that we can't leave the house without our phone for fear that we'll be out of touch. Being able to reach out and connect instantly, without the need to wait, is a dominant touchstone for our era. The seeds of this enchantment of the instant were planted back in the mid- to late nineteenth century with the launch of the telegraph and the pneumatic tube systems.

In our contemporary era, being globally connected with one another shifts our sense of time, as we are able to know of news happening around the world in an immediate way. Marshall McLuhan called our electronic media Earth's "nervous system," which connects us to distant events in instant ways.[3] This instant connection that speeds up our information not only is an emotional perception but can be felt in our physical health. Doctors report that the stress of living in this sped-up digital age has a direct impact on our heart health, as we feel obligated to be in touch and available to our jobs, our families, and our social lives at all times.[4]

If time were ever valuable, it would be at this moment in history, as we are pulled in multiple directions by our constant connectivity with the world around us and with the shifting demands on our time. A graduate student of mine recently told me about his partner, James, who was awakened throughout the night by text messages from his job. Each time his phone pinged with the notification of a message, James got out of bed to respond. My graduate student was frustrated with his partner and with the expectations to which he was responding. He made his annoyance known. James said that this was the norm in his line of work; he was expected to be available at all times.

As our technological moment combines with the shifting expectations in many lines of work, keeping people waiting is seen as unacceptable. Keeping others waiting means wasting their time. In an age when we can instantly connect with one another at any time, where the vast majority of humans on Earth are synchronized with the same atomic clock, waiting is perceived to be a result of negligence. To be kept waiting is a personal insult; people don't have much time—in fact, it feels like everyone has less time—so being forced to wait impedes an ability to use time productively.

The pull of acceleration in our technological age is simultaneously a promise and a demand. We are promised the ability to connect instantly, whether it be with the ones we love, with our work colleagues, to the latest news as it is happening, or to the facts and ideas we're seeking. The promise is grand. It offers to shrink the distance between us and the objects of our desire. Conversely, acceleration is also a demand, in that it requires a certain practice of productivity. Once we cross the threshold into the technological era of high-speed connectivity, we're responsible for keeping up. Norms shift, and what is expected of workers, of lovers, of an attentive populace is *more*.

Within both the promises and the demands of our instant culture, waiting is seen as an antiquated practice that needs to be eliminated. This is an additional promise of the technologies of an era, that they will eliminate waiting all together. Instead, time will be used efficiently and connections will be bridged. Media theorist Douglas Rushkoff has termed this "present shock." He writes, "If we could only catch up with the wave of information, we feel, we would at last be in the *now*." Yet this mythology of catching up to the present moment through our technologies is unachievable from Rushkoff's perspective: "It seems as if to digest and comprehend [these waves of information] in their totality would amount to having reality on tap, as if from a fantastic media control room capable of monitoring everything, everywhere, all at the same time."[5] This "digital omniscience," as Rushkoff terms it, is our desire but is always out of reach. Within the promise of this kind of connectivity is the pledge of a different relationship with time, one in which waiting is a flaw that is solved by the latest technological breakthrough. This was the fantasy when pneumatic tubes were laid under New York's streets in the late nineteenth century.

Molly Wright Steenson, a media theorist who studies the Parisian pneumatic tubes, has written, "If 'history passes through the sewers,' as Victor Hugo wrote in *Les Misérables*, then perhaps modernity passed through the pneumatic tubes."[6] The pneumatic tube mailing system in Paris, the *poste pneumatique*, ran from 1866 until 1984 and was one of the most extensive in the world. One of the ways that "modernity passes through the pneumatic tubes" is that the tubes present *an idea* more than they offer a solution to the problem of time and distance. In a technological culture that uses innovation and advancement as its fulcrum, the object that is raised on that fulcrum is the mythology that is thrust into the public's imagination. When a technology speeds up our ability to connect with one another, as the pneumatic tube was able to do, the words being sent through the letters in the tubes aren't the only content; as I argue throughout this book, the content is also time. It is not simply that someone received a love letter; it's also that someone got a love letter sent through the pneumatic tubes. This is the mark of an accelerated culture.[7]

Pneumatic tubes began delivering mail for the first time in London in 1853, around the same time that the railroad began connecting distant parts of England and speeding up travel times. This was a noteworthy moment in communication history because it signaled a speeding up of the transmission of messages. As Tom Standage notes in *The Victorian Internet*, until this era, sending messages took the time required by a messenger on foot or on horseback. He writes, "This unavoidable delay had remained constant for thousands of years; it was as much a fact of life for George Washington as it was for Henry VIII, Charlemagne, and Julius Caesar. As a result, the pace of life was slow."[8] It wasn't until the appearance of "something that moved faster than a horse or a ship"—a telegram, a train, mail propelled through a pneumatic tube—that communication media sped up the pace of life and peoples' connections with one another.[9]

Other cities, such as Berlin and Vienna, soon followed London to install their own pneumatic tube mail system. In New York, one of the first experiments with pneumatic tubes was a human-sized subway car that was propelled down the length of a city block by compressed air, perhaps the first ancestor of the Hyperloop currently being built in cities around the world. This experiment—which ran under Broadway between Warren and Murray Streets and was the city's first attempt at underground transportation—sat immediately across the street from City Hall and vied to become the prominent mode of transportation in New York. Ultimately, it ended up being more of an amusement ride, since the air pressure required to propel the car couldn't send it far. Regardless, the promise of pneumatic tubes was planted in the subconscious of New Yorkers.

Philadelphia, home of the Post Office Department and the foundations of the postal system in the United States, was the first city in the United States to launch a pneumatic tube mail system. The kickoff for the tubes was a memorable event. The inaugural items sent through tubes were a Bible wrapped in the American flag, a cat, a dog, and China cups; the demonstration concluded with the attendees' lunch being delivered intact. (According to a later report, a thirteen-year-old boy was sent through the tubes in a large experimental canister in Chicago.) The tubes in the United States were eight inches in diameter, which set them apart from their European counterparts, mostly three-inch tubes. The bandwidth, as we would term it, available to the U.S. Post Office Department was more than seven times that of the Europeans. That said, France had 217 miles of tubes in the country, compared with the 42 miles in the five cities that employed pneumatic tubes for mail in the United States.

Once the system was in place, people began using it to send letters and small packages to one another at an extraordinary pace. Each day, tens of thousands of notes, memos, receipts, love letters, and business propositions were sent through the tubes. The system was enormously expensive to install and maintain. Each year, Congress established a commission to look into whether the pneumatic tubes were actually worth the cost. One report from 1910 noted that it cost $120,000 per mile to install the tubes (equivalent to around $3 million dollars per mile today) and that the government paid $17,000 per mile each year to maintain them (equal today to about $423,000 per mile). At this rate, the chairman of the Committee on Post Offices and Post Roads noted, the tubes would require an appropriation of about $1 billion (equivalent today to $26 billion dollars) per year to keep this system—and all elements like workers and electrical costs—afloat. To lay this infrastructure and maintain it was not dissimilar to the scene that

Randolph Stark stumbled across in his Wall Street neighborhood—in the dot com boom era of the late 1990s, digging those trenches for fiber optic cables cost around $5.2 million per mile.

In each of these committee reports on the pneumatic tubes, tests were conducted to see whether the tubes were actually faster than other modes of message conveyance. Each year, postal workers took part in a speed test of the delivery systems available. In these tests, workers would receive two letters at a particular station, postmark them, and then send one through the tubes at the same time one was loaded onto a mail truck. The deliveries were timed. The race was on to see whether the government could find justification for spending the amount it did every year on the tubes. In 1931, a letter leaving the General Post Office, directly across the street from Penn Station, took seventeen minutes to sail through the tubes across the Brooklyn Bridge and into the receiving room at the Brooklyn Post Office. A postal vehicle that left the General Post Office arrived at the Brooklyn station forty minutes after departure. To get to Wall Street from the General Post Office again took about seventeen minutes through the pneumatic tubes; the mail truck took seventy-eight minutes to negotiate the slow crawl of Manhattan traffic.

Over the years, the delivery times for the pneumatic tubes didn't change at all; vehicles, on the other hand, got slower and slower. In the 1931 and the 1949 tests, it took a letter twelve minutes to travel through the tubes from the General Post Office to the Gracie Postal Station on 87th Street on the Upper East Side. To get there by automobile in the early 1930s took thirty-eight minutes; eighteen years later, it took an hour. Congress, which oversaw everything done by the Post Office Department, ran this test countless times over the years and produced consistent results. In each report, experts, business owners, and everyday patrons testified to the same thing: the tubes are fast, they are faster than other options, and they deliver the mail even in conditions that limit overground delivery, such as intense snowstorms.

Postmaster General Charles Emory Smith wrote in 1900, "It would not be surprising to see . . . the extension of the pneumatic tube

Figure 4. Workers at the Grand Central Station Post Office, loading the pneumatic tube carriers with mail, 1949. The open carriers are lined up next to the tube that shot the mail underground. These letters would leave the station and be timed to see whether they were fast enough compared with automobiles to justify the cost of maintaining the tubes. Image courtesy of the National Archives, Washington, DC.

system to every house, thus insuring the immediate delivery of mail as soon as it arrives in the city." Media scholar Holly Kruse notes that the prevailing idea was that eventually the tubes would also connect one house to another, so neighbors could send one another notes through a pneumatic system connecting everyone.[10] Just as fiber optic cables have come to connect everyone in the digital age, the capacity to send packets of information and messages to every house was indeed the ambition of those who ran the pneumatic tube companies of the era, as well as the dream of everyday people.

During its use in the United States from 1893 to 1953, the pneumatic tube mail system was incredibly successful. It connected people in ways that were unprecedented. While it didn't remove the need to wait (again, waiting is always a part of sending and receiving messages, no matter how fast the technology is), it did provide a new way to send messages at speeds that made people feel like they were living in the future. In this era, authors used pneumatic tubes as elements in their work. E. M. Forster, for example, employed the technology in his science fiction short story "The Machine Stops," written in 1909. In Forster's story, set in a distant future, when humankind lives underground and people rarely see each other face-to-face, the tubes supply all needs, including communication, medicine, food, and music.

But this feeling of tubes being a cutting-edge, futuristic technology faded over the decades the system was in place. In late-nineteenth-century New York, the feeling that the pneumatic tubes were the technology of the future was related more to the human experience of time than to any objective measure of how quickly they could deliver messages compared with mail delivered by car. The human perception of time shifts along with the technologies of an era and is often influenced by the cultural sense of how the pace of life is being altered by new technologies.

In other words, what it means to wait for a message changes from era to era. Waiting is experienced in its context rather than in the actual hours, minutes, or even seconds someone is kept waiting. As discussed in the previous chapter, though we are all on the same clock, how we experience that time is unique for each individual, shaped by the circumstances that contextualize waiting. The delivery of mail by the pneumatic tubes didn't change in time from year to year. Each year, it took a message four minutes to sail through the tubes from the General Post Office to Grand Central Station. It remained that way for the fifty-four years that the system was in place (minus a brief period from 1918 to 1923, when the Post Office Department, citing high costs, shut them down). Regardless, the perception of time across those decades had shifted radically. Cars had been invented and adopted at a remarkable pace; planes had been invented and were delivering mail by the 1910s. World Wars I and II had taken place, and urbanization was on the rise. The population of New York had quintupled over these years, ballooning from around 1.5 million to 7.8 million by the time the tubes were decommissioned. Everyday life looked very different in New York when the tubes were shut down for good than when they were first commissioned. Though the tubes remained constant in their ability to send messages at a certain pace, the human perception of time, along with the demands put on that system, transformed what it meant to wait for a message.

The Post Office Department gave several official reasons for the final decommissioning of the pneumatic tubes, from the ability for cars to carry massive amounts of mail of all sizes to the difficulties in fixing broken lines deep under the city streets. Yet more than for any other reason, pneumatic tubes were probably decommissioned—as is true of many technologies—due to the shifting perceptions about how well the technologies align with our expectations of time and speed. Just as notions of waiting shifted across the years in a place like New York City, so too did expectations around the "instant." Technological obsolescence is often driven by an increasing separation between notions of synchronicity (everyone being on the same clock, sharing the same information) and delay. That is, the instant is often defined in contrast to the ways we define delay.[11] What constituted instant communication in the era of the telegraph or in the era of the pneumatic tube was specific to those eras. Such notions did not hold true fifty years later, as people's perception of the instant changed with the rise of new modes of transportation and new global connectivity across two world wars.

Part of the story of the abandonment of the pneumatic tubes is one that's very familiar: technological obsolescence. Technologies get replaced by newer ones that are presented as solving the problems of the old. The tube system was set up in an era when postal workers were delivering mail by horse-drawn wagons. Though railroads were a part of this delivery process, the "last mile"—which is the industry term for the very last portion of a communication infrastructure that delivers it to the home or business of the user—was fulfilled by a slow and soon-to-be-outdated mode of transportation. Once cars made horse-drawn mail wagons obsolete and became a prominent mode of mail delivery, they changed the speed at which the last mile was achieved. Pneumatic tubes were never a "last-mile technology," and they never could be on a practical level. Instead, they allowed the mail to be shipped from one hub to another. A person still had to hand-deliver the mail or walk to the postal station to pick up the message. Soon on the heels of the installation of the pneumatic tubes came the automobile, which brought its own notions of speed (though its rapid adoption would create gridlock across Manhattan).

Once the pneumatic tube mail system was decommissioned in the 1950s, the pipes remained dormant under the city streets, out

Figure 5. A mail wagon at Grand Central Terminal, 1913. When the pneumatic tube system was introduced, it shared mail delivery with these slow wagons. This comparison supported the idea that pneumatic tubes made instant connections possible. Image courtesy of the National Archives, Washington, DC.

of view, a lost part of the history of message exchange in the United States. Randolph Stark was in his mid-twenties when he first heard about the pneumatic tubes at the technology gathering in Manhattan. He was just a few years out of his undergraduate studies in liberal arts at St. John's College in Annapolis, Maryland. He had gone into the field of engineering soon after graduation, trying to blend his humanities skills with the rapidly emerging technology field. Even though the tubes he was trying to resuscitate had been decommissioned for a half-century (three years after E. B. White published his short book about New York City), there was still a living memory of the mail tubes. Besides the post office tubes, used until the early 1950s, tubes were a part of daily life throughout the twentieth century—in department stores, banks, hospitals, and the New York Public Library. Chatting with anyone who grew up in the 1950s will likely conjure memories of standing at the counter at a department store, handing the clerk money for a purchase, and seeing that money magically sucked away with a whoosh up a tube to a different floor of the store. Sailing back with a familiar clank, the tube containing the customer's receipt and change would return for the clerk to retrieve.[12]

At public libraries in Manhattan and Brooklyn, books were delivered through pneumatic tube systems, not only allowing customers access to closed book stacks in the back of the library, but also delivering the books with a speed that resonates with our own enchantment of information access. Accessing knowledge through the tubes at an unprecedented speed gave patrons the sense that ideas were available to them through technology that reduced the amount of time they had to hunt for it. This system at the Brooklyn Public Library remained in place until the early 2000s, when the system was decommissioned and removed after many books exploded from the pressure in the tubes.

Stark saw the abandoned tubes under the streets of New York as a tool to solve a major problem for the internet. "Everyone was trying to solve the last-mile problem," Stark told me, referring to the last leg of getting internet to the everyday customer. Laying cable was one thing, but getting the cable distributed to the customers and making it faster than the competition's product was a key challenge of the dot com boom (and remains so today). The first web browsers had been introduced earlier in the decade, and already people were unwilling to wait for content over their modems. The internet gave people a similar feeling as those in the age of the pneumatic tube: the technology signaled that we were living in the future. But the slow crawl of data across the line disrupted this mystique of instant connectivity, demanding a remedy from the telecommunication companies.

The move from dial-up internet to digital subscriber line (DSL) was an early response to the last-mile challenge, as users demanded faster speeds for their increasingly large websites and digital files. DSL allowed customers to be constantly connected to the internet and to have a dedicated line whose performance would remain fairly stable regardless of how many users were accessing the network; however, DSL degraded in speed with distance from the provider's central hub, and was unusable if the customer was three miles from the internet company's signal. Stark's project addressed this geographical challenge by bringing these centers closer to users. "During the height of the dot com boom, time was the most expensive thing," he said. Any proposal that would speed up the digital age was attractive, and Stark's project was alluring to companies like Goldman Sachs, which offered to help him launch his pneumatic tube dream.

Stark was surprised to find such an extensive network of pneumatic tubes under the streets of New York. These tubes had connected nearly all of the post offices in the city early in the twentieth century, and there were post offices every twelve to fourteen blocks, to accommodate people before mail was delivered directly to the home. The system spanned beyond Manhattan through tubes that ran across the Brooklyn Bridge to connect the borough with the General Post Office in Brooklyn.

Stark's plan was to lay fiber optic cables in these tubes, and just as each post office had once served as a neighborhood hub for the delivery of pneumatic tube messages, it would now be a hub for the internet in that neighborhood. The fiber optic cables would shoot off from the post office station to the nearby apartment buildings and companies, allowing these customers to be closer to a main hub for the internet. In each of these locations, Stark would work with telecommunication companies to set up servers that would allow users to connect with the internet without needing to have these lines run too great a distance for the rapid transfer of data. Messages and data would stream to people's computers with the speed anticipated from this new system.

Stark patented the idea. Yet since their decommission in the 1950s, no one knew who owned the pneumatic tubes. After digging through the archives, Stark discovered that the company that owned the pneumatic tubes had been the target of a property tax lawsuit by the city. The company then abandoned the tubes to the city, but this transfer had been lost in the records, and the city had never used this neglected infrastructure. Stark began to negotiate with New York to obtain this right-of-way for laying fiber optic cable. The city owned the tubes, but Stark owned the idea of using the tubes for the fiber optic age. He would have to work with the city to revive this forgotten infrastructure.

Months later, the dot com bubble burst and all financing for projects evaporated. Stark noted, "The dot com crash started to happen in late 2000, starting in the telecom sector. So it didn't matter if I had gold-plated WiFi cables, no one had money to do anything. Financially it was a terrible time." Soon thereafter, the World Trade Center towers came down on 9/11, and anything having to do with New York City infrastructure and innovation came to a halt. Entrepreneurs and tech innovators were no longer given access to details about the city and its buildings; all records, maps, and schematics were removed from public access as a security measure.

As the city slowly began to emerge from the trauma of 9/11, planners began to think about how to expand the infrastructure for the increasing demand for internet access. The city maintained an exclusive contract with Verizon to oversee all fiber optic access to New York City residents. This contract dates back to the same era when the pneumatic tubes were laid, in the 1890s, when Empire City Subway (ECS) was given exclusive access to run the conduit system that held all the telephone lines. ECS eventually became a subsidiary of Verizon, and it has continued its control over the cables run throughout the city into the internet era. The advantage of sticking with Verizon was that the city wouldn't pay for the laying of the new cables needed; Verizon would cover this cost. In exchange, Verizon had a lucrative monopoly. "I was about a year too late," Stark told me.

As he walked around New York in the years following the dot com boom and bust, Stark kept an eye out for Verizon trucks laying new cable around the city. He would chat with the workers laying the cable, who would tell him that it was almost always easier laying new fiber optic cable rather than utilizing old infrastructure. The crews would dig shallow trenches or even string the cables along the sides of buildings. Stark told me, "I was shocked. I was a little taken aback. I'm coming from this perfect engineering world where lines are always straight, but out in the real world they just do what they have to do to get it from point A to point B. Having to excavate down to find something which is probably not in perfect condition to begin with, that probably requires a level of care that they don't want to do. It's easier just to take that big rotating saw that they have and cut a hole or cut a trench in the asphalt and stick their garden hose size cable with the fiber technology and just take it on down the block."

Throughout history, the technologies that survive and thrive are often the ones that are able to adapt rather than fit into the perfect engineering world that Stark imagined. Citing the rise of the automobile compared with the railroad, Stark said, "Anytime that you can be sloppy and quick and dirty, you are probably going to win." Sometimes, advances in communication technologies take shape around the possibility of developing and connecting via ad hoc networks. The unsuitability of the pneumatic tubes for ad hoc and flexible networks was yet another limitation, and Stark's plan to adapt those tubes for fiber optics faced similar drawbacks.

One site that ended up successfully redeploying a pneumatic tube system was the old Western Union Building at 60 Hudson Street in Manhattan. This site was the central location for the company's telegraph message service, much of which would be transferred in the building by pneumatic tube after a message was

Figure 6. A Verizon truck parked near the installation of fiber optic cables at the intersection of 12th Street and Broadway in New York City. The installation of these cables is often a process of simply cutting a new hole in the asphalt and finding the easiest path from point A to point B. Photo © 2017 Jason Farman.

put on paper. Tubes connected the multiple floors of the building, which was erected in 1928. After the telegraph faded in use and the telecommunications industry was deregulated in the 1960s, Western Union vacated the building in 1973. Soon after, MCI moved into the building, followed by other telecommunications companies seeking to utilize the old tubes and other infrastructure already in the building. As Andrew Blum details in *Tubes: A Journey to the Center of the Internet,* these communication companies, which would establish internet, benefited from their proximity to one another. The "meet me" rooms of 60 Hudson Street, where the servers of one company connect directly with those of another, reduce latency because the data has a very short distance to travel. "Inevitably, those networks began to connect to one another inside the building, and 60 Hudson evolved into a hub,"

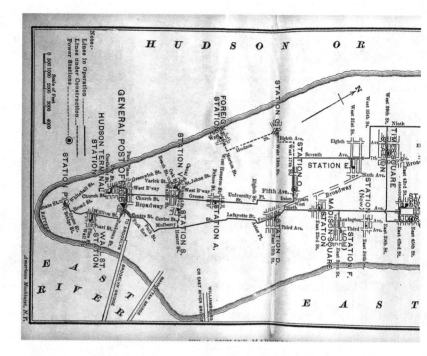

Figure 7. A map of the pneumatic tube mail system as it existed in the fall of 1909. Image courtesy of the National Archives, Washington, DC.

Blum writes. "It's the paradox of the Internet again: the elimination of distance only happens if the networks are in the same place."[13] Now 60 Hudson Street is one of the key sites for the internet globally, as the undersea internet cables that connect the United States with other countries around the world come ashore in Long Island and New Jersey, eventually meeting in this building. The cables connect servers through the pneumatic tubes before sending out their data across this part of the country. As Blum notes, this building's connection with London's Telehouse internet hub constitutes one of the busiest internet connection linkages in the world.

Redeployments of a pneumatic tube system, like the one at 60 Hudson Street, are rare and mostly out of sight of the general

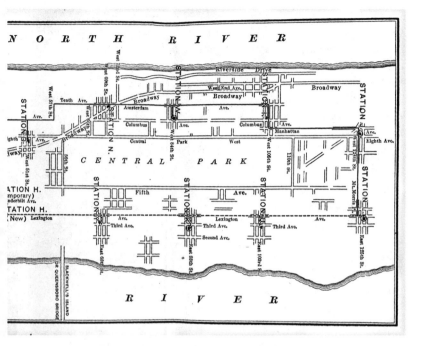

public. Most of the remains of the Postal Service's pneumatic tubes have been out of sight for decades, forgotten and neglected among the emerging infrastructures established to make life in twenty-first-century New York function.

To understand the relationship between instantaneous culture and waiting, it is important to explore the links between a specific place, a culture's perception of time, and messages in an era. I wanted to tour sites in New York where the remains of the postal pneumatic tube system could still be seen, so I visited the National Archives in Washington, DC, to gather information about key locations for the system. As I met with one of the archivists, Bill Creech, who is the National Archives' leading expert on the Postal Service, he noted that most of the archival information I was looking for had been restricted from public access since 9/11. From the government's perspective, it's entirely possible that a terrorist organization will sign up for a reader's card at the Archives to access blueprints and maps that will help carry out a terror plot in the city. Ultimately, Creech had to comb through every single box related to the pneumatic tubes—seventy-one boxes totaling nearly forty linear feet of material—to see whether any of them contained a map of the system in New York or any building schematic. These materials, if approved, were given to me as they were released and a folder would sometimes come to me with a large yellow piece of paper noting that material had been removed due to its sensitive nature.

In the end, I had to re-create the pathways of these tubes based on newspaper clippings and street addresses from the archives. Though detailed maps exist for the system, I never was given access to any of them. I came across one map from the *New York Herald* that was published just as the tubes were nearing completion in April 1887. I cross-referenced this map with the addresses listed in the extensive records of the pneumatic tubes and postal substations in New York. From these addresses and the early map, I created a new map of the tube infrastructure and began walking the length of the tubes as they ran under the city streets from post office to post office.

As I walked along the route of the pneumatic tubes in Manhattan, I came across a place that best symbolized the convergence of eras discussed in this chapter. The era of the pneumatic tube system and the era of digital technologies come together at the SoHo Apple Store, at the corner of Prince and Greene Streets. I had previously encountered this store on my way to a conference at New York University, as I made my way through police barricades and thousands of people who stood in line for the newest release of the iPhone. The line for the new device, which was launching that morning, wrapped around six city blocks. Apple uses the wait for a new iPhone as a powerful vehicle for building anticipation and desire, which paid off in the blocks of people standing for hours to get their hands on the new device.

When I came across the store in the spring of 2017, I noticed something different about the exterior of the building. Just above the front entrance hangs a metal sign with the Apple logo cut out; yet on the surface of the building, there is an old inscription, "Station A." I had actually come searching for Station A—the Prince

Street Post Office—not realizing that I would find the same Apple store from my previous encounter. In the former life of this building, it was an old post office used for pneumatic tube mail delivery. The old post office that used to send instant messages around New York was now the home to a glass staircase leading customers to the new form of instant messaging: rows of iPhones displayed on polished wooden tables. The post office that once symbolized modern life had been gutted and replaced with the new symbol of modern life and instant connection.

I toured the city, going to each site of the former pneumatic tube mail system, retracing the routes that the tubes followed through New York. Just north of SoHo Station A, the tubes went up Greene Street, about four feet underground for most of the journey. At 3rd Street, the tubes made a left turn and headed north under Washington Square Park. Many of the sites on my route are still working post offices, though most have been renovated to create a uniform and recognizable customer experience once

Figure 8. The exterior of the Apple Store in SoHo, which was once Station A or the Prince Street Post Office, from which pneumatic tube messages were sent around the city. Photo © 2017 Jason Farman.

you enter the doors. One postal driver I spoke with at the Radio City Post Office laughed when I asked him whether any of the old pneumatic tubes or pipes remained. "No, this whole place has been redone. They're even building new condos right above the station right now." The city is constantly evolving and erasing traces of its past while it updates itself.

As I left the Apple Store and walked up Greene Street, I looked at the many manhole covers and street grates leading to underground networks that keep the city working, from electrical systems to sewage, from internet cables to water lines, all out of sight.[14] Just as in E. B. White's time in the late 1940s, there's an entire ecosystem of tubes under the city, and the pneumatic tubes sat alongside these systems, all designed to be out of view. I tried to pause to notice the elements of these underground worlds that peeked up at moments to the surface of the streets, but the pace of the people walking around me made it difficult. Amid the sounds of cars, pedestrians, and construction is a silent pathway under the asphalt that once made the social life of the city vibrant and modern. As the media shift, and as our concepts of time and instant connections transform with these new technologies, this old system has disappeared. There it will remain until a crew digs up a portion of street to bury the next system that keeps the city alive and connected.

We are continually told that these systems will speed up the ways we keep in touch. There is no doubt that these technologies have indeed sped up our connections; however, we will always have a time lag that makes us wait. The notion of the "instant" that accompanies all of these emerging technologies is never fulfilled, yet it has an extraordinary impact on how we think about social connection. Of note is how we think about our wait times for messages from one another. While our technologies may be giving us a sense of an ever-accelerating pace of life, the time gap between our messages will continually be shaped by acts of waiting.

3 SPINNING IN PLACE

Before the invention of traffic signals, the flow of cars through a city was chaotic. As the automobile increased in presence in the early 1900s, it drove alongside horse-drawn carriages, trolleys, and pedestrians, all vying for passage through the snarled streets. People depended on the police officers in a busy intersection to guide traffic smoothly, and on everyone knowing the right-of-way laws. This rarely worked well and required a significant investment of labor by police forces to have multiple officers stationed at all busy intersections. On September 28, 1916, the city of Macon, Georgia, installed its first traffic signal at the corner of Second and Cherry Streets. Though not the first traffic signal in the United States, it symbolized for this region of Georgia an entrance into the modern world, where mechanical inventions would regulate the flow of traffic. Like the pneumatic tube mail system discussed in the previous chapter, the traffic signal was a sign of modernity and cosmopolitanism, putting cities like Macon in league with Paris, where traffic signals had been running with great success. However, unlike pneumatic tubes, this form of modernity slowed people down rather than sped them up. Bringing people to a halt and making them wait so that traffic could flow in an orderly way—and thus more smoothly and reliably—was the key to this next step into modern times. The *Macon Daily Telegraph* reported, "This signal is similar to the ones now in use in large cities and, according to reports, have proved satisfactory wherever they have been used, greatly reducing the number of accidents." Three days after this initial report, on October 1, the headline in the *Daily Telegraph* read, "Traffic Signal Wrecked." The reporter noted that the "dirt dug up for the new traffic signal was still fresh" when a taxi driver plowed into the new signal, forgetting that the new technology had been installed. His car's radiator plowed right into the signal, "twisted it and also twisted the 'Stop' and 'Go' hands on the traffic signal into curves."

Unlike people today (and perhaps the taxi driver who wrecked the first traffic signal in Macon in 1916), people at the turn of the twentieth century were enthusiastic about the installation of traffic signals in their cities. The waiting imposed provided a technological solution to the problem of increasing density and modes of transportation in a city. The speed of modernity required a technology that would force people to wait. A London railroad manager named John Peake Knight had introduced the first traffic light in 1868. It used a semaphore arm system, where one arm stuck out to read "Go" or "Proceed" and another read "Stop" or "Closed." A single police officer, using a series of levers, ran the mechanical system that retracted and raised the arms. At night, gas lamps illuminated green and red lenses indicating passage or the need to wait for cross traffic to pass. The first system soon came to an end after one of the gas lamps exploded onto the commanding police officer. Traffic signals like this were not used again for decades. However, the need to regulate traffic become increasingly dire with the rise of the automobile, and people all across the United States and the United Kingdom were inventing different versions of the traffic signal as a means to handle the congested streets.

Soon, electrical systems were established to run the traffic signals, with the first being installed in Cleveland, Ohio, in 1914. In this version, electric lights on each of the four corners of the intersection lit up the words "Stop" and "Move." The pace of these lights could be altered to accommodate emergencies and the need to keep people waiting at one section of the intersection longer. Today's lights have similar features, allowing cities to set up a light pattern timed—theoretically—to allow a driver going the speed limit to remain at that speed through each light on a major street. However, regulation of driving pace may be disrupted by weather, which can throw off the inner timing mechanisms of the system, or by drivers who have their own sense of how quickly they should be going down that street. The technology either fits with our rhythms of daily life or interrupts it, breaking our sense of timing and flow.[1] Every driver knows the frustration of hitting every single red light while moving through a journey.

We have an acute awareness of duration, especially as we wait, and that awareness is always linked to prevailing technologies that shape how we understand and experience time. This is true from the telegraph to the pneumatic tube system to the traffic light; each one gives us a sense of speed, connection, and the role that ideas of "bandwidth" play in how fast we can or can't go.

One such technology reshaping our contemporary sense of a moment is an otherwise unassuming little piece of interface design: the buffering icon—the animated image on our browsers spinning in place as we wait patiently for our content to load. The icon suggests that some complex code is being processed behind the scenes, as information speeds through the internet to our devices. In lieu of access to that code, we are given this animated indicator to hold our attention, and to reassure us that our request is being processed. The buffering icon's activity is meant to help us sit back and enjoy our passivity. These icons are meant to shift our expectations, modifying our willingness to wait. But the image of a buffering symbol, which has kinship with the red light at an intersection, has come mainly to trigger anxiety. As the scope of our technology use has expanded with transmission capacity, bandwidth limitations have remained a choke point, and that means that users are left waiting. As Neta Alexander has asked in her research on buffering icons, "Is buffering a punishment? And if it is, what sin have we committed?"[2]

Waiting is a part of many interface designs that we encounter. Waiting can mean vastly different things for each of these designs. For example, certain games use waiting as a core feature of play: You may have to wait an hour before your character gains another life. Some designs use waiting as an important part of how customers interact with a product, like building anticipation around the launch of a new device. Conversely, many businesses wrestle with waiting as the one factor that has potential to send their customers away. Thus waiting is a part of the design of everyday life, from managing automobile traffic to using our digital devices to create experiences for customers who encounter a business and its products.

Waiting stamps desire on one side of the coin and frustration on the other. Anticipation lives on one side, and on the other, annoyance at the inability to get behind the scenes of these systems. Waiting is a powerful tool to build our sense of connection with a new technology, and it can simplify that connection as well. The buffering icon or the simple red light gives the person one command: wait. And we wait without knowing the complexity of the larger systems in place, whether that be the code and infrastructures needed to send us an online video or the traffic patterns that allow cars to move safely without congestion in a city. Yet this ease of interaction has a cost. We often are restricted from getting to know our systems in a deep and meaningful way and are thus further detached from the inner workings of our technologies.

Again, there are times we actually prefer to wait. The times when we expect to wait can signify thoroughness and a desire that our waiting will be worth it. Waiting is a cost paid, and we want to wait in ways that will produce our desired outcomes. Waiting can represent our hopes for a positive resolution and thus waiting comes to symbolize our longing for a different future.

When the Xerox Star, among the first commercial networked computers, was released in 1981, it allowed people to do things at a speed that they hadn't been able to achieve before. The Star, officially called the Xerox 8010, connected people online and gave them the ability to exchange files and send messages to one another. Instead of using a command line as the main interface—the text-based entry system that required users to input the correct commands to make the computer carry out its tasks—the Star was one of the first computers to give users a "graphical user interface" (GUI), which remains the standard interface for our contemporary computers. Thus it sped up how people interacted with their files and with one another. Yet as Brad A. Myers, a professor of computer science at Carnegie Mellon University, told me, people felt that it was incredibly slow. Though early adopters of the computer could actually carry out tasks far more quickly than they had before the Star, the overwhelming sentiment was that it was slow at everything. People's feeling about this computer was that it took forever to load. It took forever to exchange files. It took forever to exchange messages. It took forever, even though it was faster than anything that had come before.

The Xerox Star used an hourglass cursor to indicate a processing lag. This cursor carried over to Apple's Lisa computer in 1983, Apple's foray into the GUI interface machines, which was launched a year before the famed Macintosh. The Macintosh shipped with the wristwatch icon in 1984, designed by Susan Kare, who rightfully noted that "more people had experience with a wristwatch than an hourglass." Many of Kare's designs still inspire those used on Apple computers today. A year later, though, Microsoft Windows would go back to the hourglass. Each of these "wait cursors" was static and did not animate the passage of time by spinning in place. The wristwatch wait cursor showed a perpetual 9:00 as the code loaded in the background. It wasn't until the late 1980s

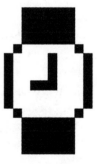

Figure 9. Wait cursors like the Windows 95 hourglass, left, and the Macintosh wristwatch make us willing to wait longer—three times as long as designs with no visualization to indicate something is happening behind the scenes.

that wait cursors started to be animated, when Unix machines introduced a black-and-white pinwheel spinning in place as the machine processed the data. It was called the "beach ball of death" because it would often spin indefinitely, until the user, finally grasping that the computer was never going to complete the process, manually restarted the machine. This wait cursor carried over into Apple's HyperCard for Macintosh. The spinning rainbow beach ball of death (the colorful version of the one on Unix machines) launched with Mac OS X in mid-2001.

The first internet version of the buffering icon was the Netscape Navigator "throbber," launched in 1994. As users waited for content to load on the browser, an animated icon would appear, with shooting stars flying around a capital letter N. Around 2006, Microsoft Vista was released; it used a circular spinning blue icon, the ancestor of what we know as the online buffering icon. Now buffering icons come in a wide range of shapes and styles, from the typical dotted circle that spins in place to website-specific loading icons customized to fit a brand.

Buffering icons and wait cursors like the beach ball of death spin in place, an apt representation of our feelings about waiting. The term "spinning in place" signifies action without progress. Phrases such as "spinning our wheels" have been used over the past century to denote actions that don't produce anything but wasted time. The twiddling thumbs that spin without creating anything or the wheels that spin in place without moving anywhere are useless. These phrases demonstrate the ways that waiting has been understood, as unproductive, a waste, since it does not propel us forward. This sentiment carries over into our feelings about the wait cursor or buffering icon. It spins and we don't know when it will stop. It is what's keeping us from moving forward and being productive. This sentiment, however, is produced because these symbols keep us from seeing how the system is actually working; we're not given a behind-the-scenes view of how the process is actually progressing, so we are kept at arm's length, spinning or twiddling our thumbs as we wait.

That said, the waiting icon produces its intended results. Waiting icons make us willing to wait longer—three times as long

Loading...

as designs with no visualization to indicate something is happening behind the scenes.[3] But we tend to respond even better when we receive some direct feedback about the progress being made behind the scenes. A better design is the "percent-done progress bar," an approach—first popularized by Myers before he started his career in academia in the early 1980s—that promises a specific end.[4] Myers began attaching percent-done progress bars to many of the tasks his company was creating for the computer. Until he introduced this feature, most lengthy processes on these machines gave little to no indication that anything was loading. There was almost no feedback about the data that was being processed behind the scenes. Myers felt that processes that would run a bit long should give this feedback to the user, and immediately customers began responding in enthusiastic ways. Despite the positive reaction to percent-done progress bars, buffering icons remain prominent. Now we have a proliferation of "indeterminate" indicators (icons that give no clear indication of when the process will finish) instead of "determinate" indicators (an icon that lets you know how much of the process has loaded and when it will finish). When I asked Myers why indeterminate indicators still predominate, he noted that conditions on the internet fluctuate widely, and a progress bar that had been moving smoothly only to stall at 99 percent is more frustrating and dissatisfying than an opaque buffering icon.

Progress bars, and the equivalent progress icons for downloading apps on a mobile device, also may have little to do with actual data-transfer rates. Designers often manipulate the visualization that purports to track download progress, front-loading it so that it moves slowly at first but then speeds up at the end.[5] This allows the download to please us by seeming to beat our expectations, which were established by the contrived slowness. Once again, technologies can establish a perception of time and duration that is independent of actual measurable seconds. Such insights have shaped the ways that Disney gives its guests feedback about wait times for ride lines at its theme parks. These wait times are almost always overestimated so that when guests reach the front of the line in ninety minutes instead of two hours, they are positively surprised that they got there sooner than predicted (rather than feeling put upon that they have had to stand in line for ninety minutes).[6] If wait times beat expectation, customers leave the experience with positive feelings rather than negative ones. On the same principle, manipulating buffering icons to beat expectations is a way that interface designers can give users a positive feeling about interacting with their software and online content.

Part of our experience of waiting is cultural, and how time elapses while we wait can vary from person to person and context to context. We wait differently and we have different expectations that are grounded in our specific cultures—from the cultural expectations about waiting in lines in Japan to a common practice in Uganda of arriving hours early to the bus stop each morning so that people can wait together as a community gathering. But while part of our perception of duration may be linked to these cultural

Figure 10. The wait time displayed at Disney's Haunted Mansion ride. Disney management typically overestimates the wait time to give people a positive feeling when they arrive before the anticipated time. Photo by Theme Park Tourist, CC-BY-2.0.

experiences of waiting, part of our awareness of duration is also a cognitive process that is wired into how our brains function. After a period of working with a particular device, according to computer scientist Ben Shneiderman, our brains begin to set expectations for how quickly it should respond.[7] If these expectations aren't met, we move on to the next task quickly (often around the two-second mark) unless something calls us back. How we wait is a combination of technological expectations (how quickly we believe that our technologies should be working), cultural expectations (how the contexts in a society set up certain expectations about how people should wait according to their position within that society), and how our brains are able to pay attention while waiting.

Though each person—depending on his or her culture—may experience the wait times differently, users across the spectrum want to feel that their waiting isn't in vain. This is especially true when users are confronted with complex systems that aren't visible, as with computer code being processed or data being sent through lines. Feedback becomes an essential tool for letting users know that the system is working, making the invisible arenas of computing life seem less threatening and off-limits.

As the mechanics of our machines recede from view behind seamless devices, we can feel detached and disengaged. Our bodies feel less connected to a machine when its systems and infrastructures aren't a visible part of our daily interactions with that machine. Allucquére Rosanne Stone traces the history of this detachment from our systems to the designs of machines in the 1930s, when the "guts" of the technologies began being hidden behind smooth and glossy surfaces. This is seen in the sleek designs of car bodies and even in vacuum cleaners. Technology often markets itself as "sleek, gleaming, seamless, efficient," Stone notes, but this is a recent experience. Previously, looking at a machine could inform you about what it did; its "affordances" were apparent just by looking at it.[8] Stone writes that the "newly constituted 'shroud,' described as streamlined, futuristic, and decorative, not only conceals the operation of the device" but creates a new relationship to the interior of the technology.[9] This interior was where things really happened, but it was a space to which people were not given easy access.

Sociologist Anthony Giddens writes of "expert systems," systems that recede from view, removing the lay person from a full understanding of how the system works.[10] Imagine, for example, that I am driving without knowing exactly how the car works, in a city where I don't know how the computers are coded that change a light from green to yellow to red. For us to have trust in these systems, users ultimately require feedback. I am willing to wait, but not if there's insufficient feedback about why I'm waiting and no information to give me a sense of control about how I wait.

Buffering icons and wait cursors confront this challenge, giving us feedback that reshapes our everyday expectations and experiences of time and duration as computers process the data being transmitted. For Myers, this was the power of the percent-done progress bar: It allowed users a sense of control over their experience. I'm willing to let a particular piece of software take eight hours to download as long as I know it's going to take that long and I can use that time in ways that I have control over (for example, I can do other tasks on my computer while the download continues in the background, or I can leave my computer altogether and do something else while this software loads). This may also reveal the allure and pleasure of multitasking; it gives us a sense that we're fully using our time instead of spinning in place. If an online movie is buffering on my browser in the background, I can jump over to email and respond to a recent message and feel that I'm not wasting my time.

That said, waiting will never be eliminated, and deep down we don't want it to be. The visible feedback about our waiting on digital interfaces should not make us think that the ultimate ideal would be to make waiting disappear. An embrace of the moments when waiting becomes visible can remind us not of the time we are losing but of the ways we can demystify the mythology of instantaneous culture and ever-accelerating paces of "real time." Notions of instantaneous culture promise that access to what we desire can be fulfilled immediately. However, this logic that dominates the current approaches to the tech industry misses the power of waiting and the embedded role it plays in our daily lives.

Buffering icons are design that gets us to wait longer than we would otherwise, and they therefore fit into a long lineage of designs meant to give us a different perception of our wait times. Famous among architects and urban designers is the story of how people stopped complaining about wait times for elevators in New York City's skyscrapers. As the *New York Times* detailed, "The idea was born during the post–World War II boom, when the spread of high-rises led to complaints about elevators."[11] The manager of one building brought in mechanical engineers and elevator companies to help him solve the problem faced every day: People were waiting too long for the elevators, and they were getting angry about it. After looking at the issue, the engineers and company representatives found the problem unsolvable. Then a psychologist who worked in the building addressed the conundrum. As one version of the story goes, "The young man had not focused on elevator performance but on the fact that people complained about waiting only a few minutes. Why, he asked himself, were they complaining about waiting for only a very short time? He concluded that the complaints were a consequence of boredom."[12] With the approval of the building's manager, the psychologist put up mirrors around the elevator waiting area so that people could look at themselves and the other people who were waiting. Waiting became interesting. Not only did the complaints cease immediately and completely, but some previous complainers actually applauded the building staff for improving the speed of the elevator service.

This example suggests that the most problematic elements of waiting are lack of feedback and boredom. The anxiety and stress caused by waiting arises from several factors. First, in a society that links productivity to time, waiting is seen as wasteful because nothing is being produced. Once people began "being on the clock"—and time was linked to salary and to the products being produced—that time was understood to be work time. Waiting in lines or on the phone for the next customer service representative is seen as wasteful because nothing is being produced that can be sold or valued:

Waiting is contrary to capitalism. Sociologist Benjamin H. Snyder has come up with two different theories to understand a capitalist approach to time. The first is the theory of regularity, which is our ability to create structured routines that help us use our time. The second is the theory of density, which demands that we pack our day full of activities that we believe are virtuous and worth the energy.[13] The result of these two approaches, however, is a population that feels overworked, busy, and burned out. We fill our time so densely and with such regularity that we end up having little time to spare.[14] In this context, waiting is a disruption.

The anecdote of the elevator also illustrates that when we are waiting, we don't feel that we have control or agency over a situation. It's not simply that we're waiting; it's also that we're not sure how or when things will be resolved. Coupled with this unease can be a deep sense of injustice. Because our time is valuable, we feel that people who cut in line or keep us from utilizing our time to the fullest are disrupting just systems where "first come, first served" is the law. In the summer of 2012, about eight miles from my house, a man was stabbed three times at a post office in Silver Spring, Maryland, because he was judged to have cut in line. The victim had been working with a postal employee to send two packages and had been asked to step aside to fill out paperwork for those packages. When he was finished, he returned to the window, as instructed. The perpetrator, who observed the customer's direct approach to the window without knowing the circumstances, went to his car to get a knife. He waited for the victim to leave the post office and stabbed him, saying, "You think you're smart? You cut in line?" The assailant's sense of justice had been violated, so he retaliated. (He received nine years in jail for the crime. The victim recovered from his wounds.)[15]

Related to the dynamic of waiting and justice is that of waiting and power. While the societal norms are clear governing waiting and lines, often those with power are able to skip the line entirely. Wealth and power often bestow immunity from the stresses of waiting, as can be seen in the "waiting economy" that is cropping up in major cities. One company called Same Ole Line Dudes, for example, will stand in line for customers to purchase tickets to the high-demand musical on Broadway or will camp out overnight to ensure that the customer will be first in line to get the latest iPhone. Other companies will stand in line for the wealthy at Department of Motor Vehicles to speed renewal of a driver's license or registration of a new vehicle.

In many circumstances, the uncertainty and precarity involved with waiting functions to reiterate the harsh divide between those with wealth and power and those without. Javier Auyero, a sociologist at the University of Texas at Austin, has studied the ways that poor people's practices of waiting often reinforce their marginal positions in a society. For six months, he studied a group of people in Buenos Aries, Argentina, as they arrived at the welfare office to pick up their checks. The results of his study of this group echoed what had been argued by theorist Pierre Bourdieu: "Waiting is one of the privileged ways of experiencing the effect of power, and the link between time and power."[16] Those who arrived at the office to receive their welfare checks would confront a scene that reiterated their position within the society. "The waiting room at the welfare office has only fifty-four plastic seats for a welfare population that far exceeds that number," Auyero writes.[17] As a result of the design of the room, many are left waiting for hours while standing or sitting on the floor. The people who come to this office take a day off work (losing that salary) and often travel long distances to arrive early to make sure they are seen before the office closes that day. People in Auyero's study reported being given the runaround, told that their money was not yet ready and they must come back another day. As a result, he argues, the urban poor of the city feel subordinated through "innumerable acts of waiting (the obverse is equally true; domination is generated anew by making others wait)."[18]

"Waiting implies submission," Bourdieu wrote.[19] While waiting is an unavoidable part of living in the world as a social being, we flee from it whenever possible because it puts us in positions of powerlessness. Yet businesses often create moments when their customers must wait. How a company's designs manage the experience of waiting can shape whether a customer avoids a business (if other options are available) or returns. Thus companies must manage customer waiting through creative designs. As the mirrors near the high-rise elevator show, our experience of duration can be readily manipulated through design, and businesses have been exploring other strategies. Some retailers are especially concerned with wait times online. As mentioned in the Introduction, an Amazon study showed that for every tenth of a second of delay that customers average on its site, it loses 1 percent of revenue.[20] Such findings have prompted the company to build servers next to those of partner companies in "co-location facilities" to cut down on the time lag, or "latency," of data exchanges. The internet hub at 60 Hudson Street discussed in the previous chapter is one such location; it repurposes the building's pneumatic tubes to connect internet servers from different companies in the building. In the mid-Atlantic region of the United States, Amazon hosts its cloud in facilities located in Ashburn, Virginia, where companies can connect to one another by simply linking their servers, some of which are just a few feet apart. This proximity reduces the time it takes to send and receive data between servers and thus reduces lag time when a customer tries to access online content. One core area of online content whose management is hyperaware of the problem of latency is online video and streaming. According to one study, after five seconds of buffering, 20 percent of people who started to watch a video will leave; after ten seconds, half will be gone. After twenty seconds, it's up to 70 percent.[21]

Here the buffering icon serves users in similar ways to the mirrors that surround an elevator. It holds our attention for a bit longer by giving us something to look at that feels like feedback. The buffering icon's minimal feedback seems to be enough to hold our attention. While it spins in place, we will spin our wheels for a handful of seconds before moving on.

Ultimately, the reasons that the buffering icon is sometimes experienced as torture are linked to all of these examples. It is indeterminate, giving us a sense of uncertainty about when the buffering will end. It robs us of our agency, because if we knew the video would be buffering for three minutes, we could leave the browser and do something else while the video loaded. The buffering icon is an opaque system that doesn't give us a behind-the-scenes understanding of what's happening, so we end up feeling a sense of detachment from our technology. Yet the icon offers enough distraction to keep us around a bit longer, like the people standing on the ground floor waiting for the elevator to arrive, staring into the mirrors.

There are other circumstances when we prefer to wait. In 2016 Facebook began offering security scans of user profiles, sending back details of any potential threats it could detect from users' profile settings. Facebook could conduct these scans rapidly and at first would spit back the information instantly to users. When it did, people didn't trust it. They didn't believe the scan was thorough, and many declined to change their settings. When Facebook inserted a bit of code that made the system pause, people began to trust the results more and make changes to their profile's security settings.[22] Travel sites also modify the speed of results, building in a false latency to try to make consumers feel the searches are more thorough. Technology is actually ahead of our expectations, yet our temporal expectations of thoroughness dictate how we experience it and the code that is written for it.

Similarly, the desire for waiting is built into launch events, as when Apple implements anticipation as a core feature of new products. Apple announces a product and makes us wait, building our desire. The company recognizes the power of having the customer's imagination at work during that wait time. These launch events themselves are pitched months in advance, and the excitement builds as images and specs of the new device are released. Leaks are often exploited to feed the anticipation of these launch events; for example, previously unreleased photos of the upcoming iPhone are routinely given to tech websites long before Apple reveals details about the new phone. On the day a new iPhone is released, people stand in long lines outside of Apple Stores all around the world. Waiting for the release leads to waiting in long lines, sometimes camping overnight, to transform the release of a device into an event that builds anticipation and customer connection to the product.

In *A Lover's Discourse*, Roland Barthes describes the eroticism of waiting. He writes, "Waiting is an enchantment: I have received orders not to move. . . . Am I in love?—Yes, since I am waiting. The other never waits. Sometimes I want to play the part of the one who doesn't wait; I try to busy myself elsewhere, to arrive late; but I always lose at this game: whatever I do, I find myself there, with nothing to do, punctual, even ahead of time. The lover's fatal identity is precisely: *I am the one who waits*."[23] The lover is not only willing to wait for the object of desire but is defined by that willingness to wait. While we wait, our desire grows and comes to define our relationship to the person (or object) we long for.

Sometimes the waiting is the very act that gives us pleasure in these erotic connections to people and things. Barthes recounts a Chinese tale of a man in love with a courtesan, who tells him, "I shall be yours when you have spent a hundred nights waiting for me." On the ninety-ninth night, the man "stood up, put his stool under his arm, and went away."[24] Waiting was the practice of dwelling in the fantasy about the object of longing.

Figure 11. Customers wait in line for the release of a new Apple iPhone in France, 2012. Launch events use waiting as a way to build desire for a new product, creating our connection with this new commodity. Photo by Patrick Hertzog/AFP/Getty Images.

Waiting is such a powerful part of our relationships—with people we long for, with objects like iPhones that we may long for—because that's where imagination does its work. For consumers and users of contemporary technology, waiting is deeply connected to our fantasies about who we are and what our purchases say about us. This was famously detailed by sociologist Colin Campbell in *The Romantic Ethic and the Spirit of Modern Consumerism*. Campbell argued that modern consumers shape their identities by fantasizing about how a product will lead to the lifestyle they are daydreaming about, what he calls "autonomous, self-illusory hedonism."[25] For Barthes, the same is true of how we wait for the ones we are in love with and long for. When the thing longed for finally arrives, it can rarely live up to the excitement generated by our imaginations.

Though it is counterintuitive, a similar logic is at play in our online lives. For me, in my moments of boredom, as I turn to my phone and refresh my social media feed, I imagine that what's on the other side of the buffering icon might be the content that will rid me of boredom and produce a satisfying social connection. The buffering icon here represents my hopes for the many ways that my social media feeds can satisfy my longings at any given moment. They rarely do, though I believe that we are half in love with the buffering icon because it represents the promise of intimacy or excitement across the distances that separate us.

Buffering shapes the relationship between duration and desire. Similar to one of the issues pneumatic tubes faced, the buffering of online content is a result of bandwidth limitations. Online content is subject to buffering as its scale increases, as more and more messages are sent across the lines in higher file sizes. The buffering icon has become one object among many that shapes our experience of time. It weaves itself into our daily lives and signals a patience that is required for what we desire. In the end, it can simultaneously create a feeling of helplessness due to the lack of feedback and stoke our desire for content that will satiate our needs. Buffering presents us with the promise of living without boredom, of connecting to our loved ones in deep and meaningful ways. Buffering symbolizes the distraction from spinning in place that we feel as we wait in the digital age.

4 SPACE SIGNALS

Hal Weaver, the project scientist for the New Horizons space mission, pointed to the image on the auditorium screen as he said, "Up until now, this is what we knew about Pluto." The image was an incredibly blurry series of white pixels against a black background. He was speaking to the Astronomy Department at the University of Maryland in 2016, a little more than a year after the New Horizons spacecraft made its closest approach to Pluto. The Hubble telescope had taken this pixelated image of the distant planet twenty-two years earlier. What became known after the New Horizons flyby of the planet changed everything for these scientists. The very last trickles of data from New Horizons had arrived back to the Johns Hopkins Applied Physics Laboratory (APL) in Maryland the day before Weaver gave his talk. It took sixteen months after the spacecraft's closest approach to Pluto for all of New Horizons' data to arrive back on Earth. During its flyby of Pluto, the spacecraft took high-resolution images, captured data about the surface and its atmosphere, and gathered images and information about its moons. Once the flyby was complete, the spacecraft turned and faced Earth and beamed its data across the expanse of the solar system. What was learned from this data has changed how we think of planets and the solar system. Weaver showed a picture of a U.S. postage stamp from 1990 with a painting of Pluto and the caption, "Pluto: Not

Figure 12. Hal Weaver, the project scientist for the New Horizons mission, display-ing an image of Pluto taken by the Hubble telescope in 1994. Until the flyby of the *New Horizons* spacecraft, little was known about the dwarf planet. Image © 2016 Jason Farman.

Yet Explored." Weaver responded, "We have made this stamp obsolete."

The data from New Horizons are shot back to Earth via radio waves from the dish antenna on the spacecraft. These are faint messages, sent from three billion miles away, and it takes enor-mous satellite dishes strategically positioned at three locations around the globe to capture these whispers of data. The mes-sages arrive in a slow drip; a far cry from the torrential flow of data on a home broadband connection, as people stream movies or music. The New Horizons data accumulate over months, with the most critical data arriving first. For the New Horizons mission, data were downloaded ("downlinked") from the spacecraft at the rate of 2,000 bits—0.00025 megabytes—per second. This is a trickle of data compared to the U.S. national average download speed on home computers in 2017, which was more than 70 megabytes per second.

The New Horizons mission is a perfect example of the vital relationship between waiting and knowledge. The unknown creates speculation as we try to fill in the gaps of knowledge with everything from educated guesses to fear-inspired myths about what lies beyond the edge of our understanding.

This mode of speculation creates a new way of thinking. Our imaginations allow us to access that which does not yet exist and create scenarios that have not yet happened. Wait times are key to this mode of creative thinking because they afford us the opportunity to imagine and speculate about worlds beyond our own immediate places and speculate about the possible.

This capacity stands in stark contrast to the ways that knowledge production has been described in the digital age, when any question or topic of curiosity can be resolved through a search engine, and our thirst for answers is quickly satiated. While this instant gratification gives us immense pleasure, such instant connection to knowledge may be transforming who we are and the neural mappings in our brains. Technology writer Nicholas Carr has argued, controversially, that the internet has fundamentally changed how we think and act. Media like the internet "supply the stuff of thought, but they also shape the process of thought. And what the Net seems to be doing is chipping away my capacity for concentration and contemplation."[1] The internet, with its ability to create links across all information in an immediate way, ultimately creates a new way of accessing knowledge and thinking. The linearity of previous media (from books to films) shaped certain ways of thinking, and thus formed the very nature of the ideas we created. Linearity, as a way of engaging media, created a specific mode of thinking and knowledge production. Some have called this mode of learning and knowledge production a "deep dive"—we spend a significant amount of time diving to the depths of an idea or field of study. It takes time and focus to learn the contours and nuances of an idea, its history, and how it was situated in specific cultures. According to Carr, these deep dives contrast the kind of hyperlinked learning, quick skimming of information, and knowledge production that happens in the digital age. He writes, "Once I was a scuba diver in the sea of words. Now I zip along the surface like a guy on a Jet Ski."

He goes on to mark this distinction with an anecdote about Samuel Johnson, a prolific English writer, and his visit to the house of the poet Richard Owen Cambridge. A year before the United States would declare independence, Johnson entered Cambridge's house and began reading the spines of the books in his library. Cambridge replied, "Dr. Johnson, it seems odd that one should have such a desire to look at the backs of books." Johnson responded, "Sir, the reason is very plain. Knowledge is of two kinds. We know a subject ourselves, or we know where we can find information upon it." From this story Carr reflects, "The Net grants us instant access to a library of information unprecedented in its size and scope and makes it easy for us to sort through that library—to find, if not exactly what we are looking for, at least something sufficient for our immediate purposes." He continues, "What the Net diminishes is Johnson's primary kind of knowledge: the ability to know, in depth, a subject for ourselves, to construct within our own minds the rich and idiosyncratic set of connections that give rise to a singular intelligence."[2]

In contrast to these losses of focus and deep dives, digital technologies create interactions that free portions of our brains' functions to focus instead on new things, in much the same way that writing freed us from the need to store every important fact in our memories. Many of us no longer remember phone numbers, because they are programmed into our mobile phones. While this puts us at a disadvantage when our battery dies and we need to access a number, the ability to liberate functioning parts of our memories allows us to use that brain capacity for other uses. Media scholar Walter J. Ong points to this principle when discussing the transition from oral culture to literate culture, noting that mnemonics were essential for remembering facts and storing knowledge before the written word could preserve these memories beyond the life of a single individual. He goes so far as to say that certain modes of complex thought would never be engaged because they would just as quickly be forgotten without the aid of the written word. He writes, "In an oral culture, to think through something in nonformulaic, non-patterned, non-mnemonic terms, even if it were possible, would be a waste of time, for such thought, once worked through, could never be recovered with any effectiveness, as it could be with the aid of writing. It would not be abiding knowledge but simply passing thought, however complex."[3] The internet and digital technologies may be freeing us to similarly engage with more complex ways of thinking now that our brains are not occupied with the need for older forms of learning and memory. Thus while a deep dive has its advantages, so too does the ability to piece together different areas of knowledge to form something new. New forms of knowledge are being produced as teams of people can simultaneously collaborate, even at different spots around the globe, to crowdsource solutions to previously unsolvable puzzles. For example, online players of the game Foldit were able to collaboratively solve the crystal structure of the Mason-Pfizer monkey virus retroviral protease in three weeks; scientists had been trying to solve this problem for a decade.[4]

While Carr argues that the internet's threat to how we think is found in its nonlinearity, the greater risk comes from the instant gratification of finding immediate knowledge that does not incorporate the necessary pauses for us to process and innovate on the ideas that we find. Efficiency here is the threat to innovation, where waiting, delays, and gaps between ideas allow us to come up with new ways of thinking. Waiting is necessary for learning certain things about the world and universe. It is also necessary for certain ways of encountering knowledge, which require delays for information processing and methods for learning something over a longer period of time. Waiting, ultimately, is essential for imagining that which does not yet exist and innovating on the knowledge we encounter.

Scientific explorations, such as those from the New Horizons team, work very differently and exist on a completely different spectrum of time from the instant connection to knowledge on the internet. To risk overstating the obvious, scientific knowledge—especially in the fields of space exploration and astrophysics—takes time. Waiting is typically a necessity, but can also be a strategy.

A mission like New Horizons can take decades from the planning and proposal stage, to the building of the spacecraft, to the launch, to the long distance the spacecraft must travel. Alice Bowman described to me the pacing of a program like New Horizons. Bowman ran the New Horizons show as the first woman to hold the position of mission operations manager (MOM) on a space mission. We stood together just outside of the Mission Operations Center on the Applied Physics Lab (APL) campus in suburban Maryland, about twenty miles north of downtown Washington, DC. Large glass windows looked in on a calm day at the Ops Center, with a single person sitting at a computer viewing green code on his screen. The lights flashing above him indicated the success of the current uplink with the spacecraft. The center was sending data, carried by a radio wave signal from a massive radio antenna in the Mojave Desert in California. The technician was also on his laptop, looking at Second Amendment websites while simultaneously scanning the uplink with the spacecraft on the desktop computer. He was surrounded by rows of computer stations, all unmanned on this winter day. At the front of the room were massive screens filled with information about the spacecraft. The layout of the Mission Operations Center—from the lighting to the kinds of chairs used—is exactly what you'd imagine an Ops Center to look like.

As we looked over this scene, Bowman noted that since 1989, Alan Stern (the mission's primary investigator) has worked to get a mission going to Pluto. The proposal for the mission was submitted and New Horizons was funded in 2002. Funding did not come easily. In the years between 1989 and 2002, several proposals for a Pluto mission were submitted, but later abandoned due to either cost restraints or technological hurdles. The choices reflected NASA's changing priorities. Alan Stern, who had finished his doctorate in 1989—the same year he began lobbying for a Pluto mission—rebelled along with other young scientists. Stern argued that a trip to Pluto was urgent. First, the planet's "atmosphere might 'snow out' as it moved into a more distant part of its orbit," as Michael J. Neufeld, senior curator at the Smithsonian's Air and Space Museum, writes.[5] He continues, "Every passing year meant more of the planet going into a decades-long night." Second, in order to reach Pluto within nine years, the spacecraft would need a "gravity assist" from Jupiter; in other words, while orbiting Jupiter it would gain a speed boost from that massive planet, enabling the operators to slingshot the probe to the farthest reaches of the solar system. To reach Jupiter for the gravity assist, the mission would need to launch by 2006. The next window of opportunity for a gravity assist wouldn't come again for another decade.

NASA was persuaded and moved forward by announcing a competition for a Pluto mission. Stern's team at the APL put together a compelling proposal, but the Bush administration, after taking office in 2001, canceled the Pluto mission in favor of other missions that had more political backing. In response, Maryland senator Barbara Mikulski put pressure on NASA to move forward with the competition, which was awarded to Stern's team in late 2001. When the program was canceled again a year later, the result of a cut in the 2002 federal budget, Mikulski interceded a second time by extending the budget to include the mission. New Horizons finally had funding to move forward.

Bowman commented to me as we looked at the green lights flashing in the Ops Center, "Then we had to build the spacecraft. It took us four years, which sounds like a long time, but it was short in terms of spacecraft. So there was a four-year wait just to get it going. And then we launched in 2006." New Horizons was the fastest object ever to leave Earth's atmosphere. "But then there is this wait from 2006 to 2015 while we're trying to get there. Then there's all the stuff that happens in between," she said. In part, she was referring to the reclassification of Pluto from planet to a so-called dwarf planet. This reclassification occurred in August 2006, just eight months into the New Horizons mission. Bowman believes that the controversy actually helped the mission gain public support: "People who were not involved with science at all had learned in grade school there were nine planets. One was taken away, and everyone had an opinion." The pro-Pluto sentiment, Bowman says, added to the anticipation of the space-craft's approach: "People wondered, 'Are we going to see a gray lifeless rock when we go to Pluto?' What we saw were mountains and glaciers and atmosphere and color. It was amazing. It still is amazing!"

During this journey, communication with the New Horizons spacecraft required longer wait times between sending and receiving messages. As Bowman noted, the farther the spacecraft gets from Earth, the time it takes to send and receive commands gets longer and longer. This means that responding to the spacecraft takes extensive moments of waiting; moments that get longer as the spacecraft travels farther into space. When New Horizons arrived at Pluto, after nine years traveling across our solar system, it was around three billion miles away from Earth.

But that arrival wasn't the end of the story for the scientists. They had to wait as messages were downlinked from the spacecraft back to massive satellite dish antennas on Earth. Though radio waves travel at the speed of light, it still took nearly four and a half hours for the bits of data to arrive.[6] And they did arrive in "bits"; slowly, the messages, images, and data accumulated as they made the long journey of three billion miles.

From the many years it took to get the Pluto mission approved, to the long journey to arrive at the planet, to the many months it took just to downlink data from the spacecraft, space exploration creates a certain pace based on the scale of the project. This expansive scale is unique in a couple of ways. A person's career in this field of study might encompass only one or two missions, because each takes so much time. So the contributions of a single individual span the many decades needed to put into a space mission like New Horizons. As scientists build knowledge, they do so on a timescale that limits the amount of knowledge they can produce.

Yet when a project like New Horizons spans the universe, the time it takes to get there transforms the everyday person's sense of scale and place in the universe. For me, as a scholar of digital media who is used to accessing information instantly and incorporating it into my research just as quickly, the idea of waiting decades to get to my primary sources is staggering. The sixteen-month downlink time alone captured my attention and shifted my own sense about time and the creation of knowledge. For many who were similarly fascinated by the mission to Pluto, the timescale emphasized the distance between Earth and this distant dwarf planet in our solar system. Wait times help concretize this kind of scale, which is difficult to fathom otherwise.

Though scholars of digital media might be in awe of the kind of wait times required for knowledge to be gained about space, this relationship with time probably resonates with many sociologists. One method sociologists use to study the world is called a longitudinal study. Instead of simply studying a segment of society and making observations about that culture, social scientists study an aspect about a society many times across many years to understand change within that culture. One famous longitudinal study began in 1938 on Harvard's campus, focusing on the health and happiness of 268 of its students. The study, which is still ongoing today, eventually incorporated the students' offspring and those of the control group of inner-city Boston youth. Included in the original Harvard students was a young John F. Kennedy, twenty-three years before he became president of the United States. In 2018, almost eighty years after the study began, nineteen of the original participants are still alive. The years of data from this group and their offspring enabled social scientists to point to a vital factor in the health and happiness of these individuals. As Robert J. Waldinger, the director of the project at Harvard Medical School, has written, this study reveals that healthy social ties are a key to health later in life. For women, strong ties and attachment security "predicted better memory . . . and attenuated the link between frequency of marital conflict and memory deficits."[7] He concludes from this study that as we age, our health is not simply a matter of paying attention to our bodies and making healthy decisions about diet and exercise. Instead, our social ties to strong relationships are just as important to living a long and healthy life.

What these examples reveal is that there are some things that cannot be known without wait times. In an era of instant knowledge, when immediate access to facts and ideas has come to be expected, it is sobering to recognize the kinds of timescales that many kinds of knowledge still require. These emphasize both the finite nature of an individual's time on Earth (as illustrated by the lengthy process to get the mission to Pluto off the ground and by the duration of the longitudinal study of the Harvard students) and our place in the universe. The passage of time creates a sense of space that, coupled with the new knowledge we have about the previously unknown details about Pluto, changes how we think about ourselves in this solar system, and about what actually constitutes the solar system.

After its Pluto flyby, the next object in the New Horizons mission is an object in the Kuiper Belt. The Kuiper Belt is the outermost part of our solar system, which rarely makes it into a young person's model of the planets or an elementary school's diagram

Figure 13. One of the first images released after the flyby of Pluto on July 14, 2015. The image surprised many with its diversity of landscape, especially the area called Sputnik Planitia, the heart-shaped lighter area in the lower right. Image courtesy of NASA/Johns Hopkins University Applied Physics Laboratory/Southwest Research Institute.

of the solar system. New Horizons' flyby of Pluto and journey into the Kuiper Belt will transform how we imagine the solar system.

In the summer of 2013, the many stakeholders for the Pluto mission came together for a conference on the APL campus. The meeting's driving question: What will New Horizons find? Experts at the conference worried the answer would be disappointing. "Many thought that since the planet was so distant, cold, and small, there likely wouldn't be any terrain diversity or signs of tectonic movement on the planet," Hal Weaver explained.

New Horizons' first images showed a Pluto that looked very different from what scientists had expected. They found that Pluto had a very young surface that was continually regenerating in a specific region called Sputnik Planitia, otherwise nicknamed The Heart. Bowman described the process of discovery: "So what you gained over time was this deeper understanding . . . of the amazing things that could cause something like this to happen." She outlined both the mysteries that scientists had hoped New Horizons would solve, and the new enigmas introduced by the spacecraft's data.

Speculation and anticipation are keys for exploration and discovery. When the Apollo 8 mission sent humans around the moon for the first time on Christmas Eve, 1968, the spacecraft lost communication with Houston as it dipped out of view onto the far side of the lunar surface. In the sixty-eighth hour of the mission, Apollo 8 began its lunar orbit. Astronaut Bill Anders radioed to Houston, "Thanks a lot, crew. See you on the other side." They then went into radio blackout. The three astronauts onboard were on the dark side of the moon and unable to use radio communication for more than half an hour. Many worried that if anything went wrong, the spacecraft would either be pulled into the moon's gravity and crash or miss the gravitational pull and be shot out into space with no ability to return to Earth.

During this time, millions of people watched for the astronauts to reemerge on their television screens. It was the most-watched television broadcast ever—and it was a broadcast of waiting.

Clayton Anderson, now an astronaut himself, was nine years old at the time and remembers vividly how he felt while waiting for Apollo 8 to come back from the dark side of the moon. "I was almost panicked by the astronauts' loss of communication with leaders in Mission Control. All I could think of during the excruciatingly long minutes of silence was that some disaster had befallen the brave crew. Imagination took over my youthful brain and I wondered if a lunar volcano had erupted into their orbit and burned them to bits. Or perhaps an evil space dragon had awakened from its slumber on the dark side and had blasted the spaceship with its fiery breath!"[8] Waiting causes the imagination to run wild, especially when we're faced with a set of circumstances that are beyond the scope of our knowledge. Since our moon is locked in rotation with Earth, humans had never fully seen its dark side before this mission, so it was an unknown territory. One NASA crewmember in Houston broke the silence while they were waiting by joking, "What about that mountain we didn't know about that's higher than the orbit of the spacecraft? Aren't they going to smash into that thing?" Thirty-five minutes later, the sounds of the Apollo 8 radio signal returned to the radio antenna on Earth—at the same site in the Mojave Desert in California, called Goldstone, that would wait for radio signals from New Horizons nearly half a century later.

When we don't know the answers to questions about how life and the universe function, we send explorers out into unknown territories to find the answers. From nautical explorers of the fifteenth century to early frontiersmen exploring the United States to the Apollo 8 astronauts, such trailblazers have inspired cultures to speculate much as the nine-year-old Clayton Anderson did about fantastic possibilities. Before explorers were sent out across uncharted oceans, medieval maps cautioned of some of these unknown areas, "Here Be Dragons." This warning simultaneously created fear and sparked the imagination about the unknown.[9] As societies have waited for the return of their explorers, they narrated the many possibilities—they theorized and speculated—in order to deal with the anxiety caused by the unknown and its link with time and the vast distances that separated them from the explorers.

These moments of silence are important to understand because they highlight the interpretive process that is a part of how we communicate. Similar to the theory discussed in chapter 1 that the spaces between letters in a word are just as essential as letters themselves for communicating meaning, silences and gaps function to help create meaning. This interpretive process, as we speculate and fill in the gaps, is not simply a side effect of ignorance; instead, it's essential for knowledge building. Speculation, built on the foundations of silence about the unknown, allows our creative capacities to build worlds and problem-solve within those imagined environments. Waiting, as represented by silences, gaps, and distance, allows us the capacity to imagine that which does not yet exist and, ultimately, innovate into those new worlds as our knowledge expands.

Lisa Messeri, an anthropologist who writes about the ways that space scientists transform the planets they study into meaningful places, notes that we are continually framing any planet in reference to our own. This approach makes sense, but has drawbacks. As we conceive of space missions and imagine distant planets, we consistently use the "frontier" metaphor of the American West as a framing device to help us understand space missions (and their value). By using this metaphor of the frontier, and by thinking of all the ways these planets may relate to Earth, Messeri argues, we actually do a disservice to the complicated ways that space exploration expands the scope of our knowledge. She writes, "The commercial space industry prides itself on newness and novelty, and yet the reliance on the same old metaphor both limits the imagination of humans in space and glosses over the social and historical problems of imagining a frontier that is empty and beckoning."[10]

Building on Messeri's ideas, we see the value that waiting brings: It gets us to imagine that which is beyond the scope of what we know in the present. It gets us to imagine other kinds of encounters and other possibilities. Waiting, along with the speculation it creates, gives us the capacity for new worlds that far exceed the limits imposed by our own. Jane McGonigal, theorist of digital games, has argued for the role of the imagination in creating new and better worlds. As director of game research and development with the Institute of the Future, she has argued, "In order to make a change in the world or invent something new, you have to be able to imagine how things can be different. And the future is a place where everything can be different." She goes on to say, "Thinking about the future shouldn't be about trying to be correct. It should be about trying to be creative, and surprising. If you imagine many possible futures, you open up the space for innovation, creativity and change."[11] Within these new spaces created by the speculation and anticipation introduced in times of waiting, we are able to create scenarios that expand our ideas of what's possible. For many artists and creators, only within the realm of the imagination produced by speculation can innovation exist.

The practice of speculation is an everyday part of the work that scientists do at places like NASA, APL, and the Jet Propulsion Laboratory (JPL), all collaborators on New Horizons. I sat down with artist Steve Gribben in his office at APL; he is the main artist and graphic designer for the laboratory and is responsible for turning speculations into images and videos of planets that have not yet been visited. He sat behind several connected monitors displaying various pictures of Pluto and the software he uses to visualize these planets. Steve is typically given some broad directives about what the planet may look like and asked to produce a visualization that will spark the imaginations of the public and scientists alike. Before New Horizons did its flyby, any artistic impressions of Pluto were based on a few details known about the planet: It had an atmosphere, it was reddish in color, and it had a moon. New moons would be discovered before the launch and during the trip to Pluto. "For the public, I want something dramatic. It gets their interests, but at the same time I have to satisfy the scientists because they'll give me feedback, too." For Gribben, his artwork "gets the general public interested. It's something to look at in awe."

Gribben's creations serve to represent the ways that scientists are thinking about the unknown. They simultaneously create excitement and engagement from the larger public, which sees

Figure 14. Steve Gribben's concept of the New Horizons spacecraft as it approached Pluto and its largest moon, Charon, in July 2015. Image courtesy of Johns Hopkins University Applied Physics Laboratory/Southwest Research Institute (JHUAPL/SwRI).

an actualization of things unknown. Representation shapes the ways that we respond through the speculation it encourages. Beyond some basic parameters, Gribben is free to depict the planets as he sees fit. He has no anxiety about "getting it right." Instead, he says, "I always go back to artistic license. The only thing I sweat about is the quality. I want it to look as sharp as possible, the colors, the saturation, all of the technical stuff. They tell me sometimes, 'We really don't know what this is going to look like, so make it look it good.'" Gribben had spent years speculating with the mission scientists about what Pluto and its moons would look like. Once the first images from New Horizons started to come in, they revealed that Gribben had gotten right some features that the mission scientists didn't. When he was asked to create an image of what Pluto's atmosphere would look like, he went with a bluish color because of its distance from the Sun. The scientists asked him to correct the colors to brownish or pale. In the end, New Horizons photographs of the atmosphere revealed a rich blue color to the atmosphere. Ultimately, Gribben's goal is to inspire people to envision the possible as a mission is ramping up or is nearing a critical moment.

In the months before New Horizons' closest approach to Pluto, NASA and JPL released a series of free posters to the public, the Exoplanet Travel Bureau, featuring imagined tourism posters of the future to far-off planets like Trappist-1e. A year later, they released another series of posters called Visions of the Future, which depicted a future when tours to historic sites of Mars would be available long after the planet had been colonized by humans. Members of the six-person design team at JPL imagine their work as helping mission scientists develop future scenarios—as JPL designer David Delgado says, "Help them think through their thinking."[12]

Scientists at NASA, JPL, and APL are all keenly aware of the impact such campaigns have on public engagement and the importance of this kind of engagement for the work that they do. Bowman called New Horizons, funded by taxpayer dollars, "everyone's mission." As such, the kinds of missions that generate enthusiasm from the broader public are the ones that get funding.

Figure 15. Jet Propulsion Lab's Exoplanet Travel Bureau poster for Trappist-1e. The caption for the poster links an imaginary future with current space missions by noting, "This system was revealed by the TRAnsiting Planets and Planetesimals Small Telescope (TRAPPIST) and NASA's Spitzer Space Telescope. The planets are also excellent targets for NASA's James Webb Space Telescope." Image courtesy of NASA/JPL.

A key factor in these missions getting off the ground in the first place is public support. And an essential component of public support is generating public excitement about the possibilities that these missions may uncover. The new knowledges that may be encountered in exploration build the anticipation that is an essential element for the public's connection to these missions.

Waiting is necessary for knowledge production because of the amount of time certain explorations and research take, as well as the ways that waiting encourages speculation and public engagement. Beyond these conditions, there are certain *ways of knowing* that require wait times. The wait times that contribute to certain ways of knowing create what are called "enabling constraints," which allow innovation on the knowledge obtained. Knowledge requires waiting in part because of the ways that we process this information and transform it into accessible, long-term memories.

One of the major constraints for the New Horizons mission is the distance between Earth and the spacecraft. This constraint creates challenges that must be planned for years in advance. The communication lag—approaching four and a half hours, as the New Horizons spacecraft gets farther away from Earth—between the team at APL and the craft makes troubleshooting problems a challenge. Ideas about what might be going wrong with the spacecraft can't be instantly tested; instead, the team has to create a contingency list of protocols for possible solutions.

The time lag between the spacecraft and the operations center at APL created a memorable moment of crisis that put the mission into jeopardy on July 4, 2015. Chris Hersman, the chief technical officer of the mission, was at a family barbecue celebrating Independence Day when he got a call that the spacecraft wasn't responding. It had gone silent at a critical moment, just ten days before its closest approach to Pluto. Before Hersman got the phone call, the team had been slated to send the commands to the spacecraft to initiate its final approach to the planet. In the journey up to this point, the spacecraft had been "hibernating" at moments across the expanse, and as it started to get close to the Pluto, it was capturing data about the atmosphere to determine whether drifting dust and debris posed any threats to the instruments on board. Now, as the spacecraft began its approach, the team sent a set of commands to New Horizons to initiate the various elements of its engagement with the planet and its moons. As the team was about to uplink the new directives for closest approach, the spacecraft rebooted, went into a safe mode, and became uncommunicative.

Alice Bowman called Hersman from the Missions Operations Center at the APL campus in Maryland. "She wasn't calling me from her office line. When she calls from the Mission Ops room, it's a bad sign," Hersman told me. "Can you tell me what's wrong?" he asked. "We've lost telemetry," she responded. The loss of telemetry—the spacecraft's communication of data to Earth—couldn't have happened at a worse time. Compounding the problem, Bowman knew that something was different about this silence from New Horizons. As the team had started the various steps to send the new program that morning, everything had been working fine. Then, without warning, there was silence. Everything on the ground was working perfectly, so the problem had to be with the spacecraft. The operations team was called in to troubleshoot, and Hersman left the family barbecue and drove to Mission Control at APL to find out why New Horizons had stopped communicating with Earth.

Troubleshooting a problem with a scientific instrument is one thing if a scientist is able to test solutions in real time. However, any commands sent to New Horizons on this day in July would take around four and a half hours to send and another four and a half hours to receive verification from the spacecraft. The team did not have much time; there was a window of three days before New Horizons would reach the point where the program wouldn't work, and the team would be unable to capture all of the scientific data it had planned for during for the flyby.

As Hersman pulled into the APL campus, the parking lot was mostly empty, as employees were spending the holiday with their families. As Hersman walked into the room, Bowman was noticeably troubled. Bowman considered New Horizons a part of herself, and she felt a meaningful link with this object flying through space. Now that link had been severed for some reason. The mission to get to Pluto, which would happen in just over a week, had been in the works for twenty-six years and was now in jeopardy of failing to gather the ideal set of data and images that generations of scientists had sought to retrieve. The team had to figure out how to bridge that silence and reestablish communication with New Horizons.

Bowman has said that waiting is a core element for how she approaches problem solving: "Your first reaction is always to try to fix things right away . . . but sometimes, waiting for the spacecraft to take care of itself for a certain period of time—like it's programmed to—is what you need to do." So instead of sending commands to initiate what they thought would be the best solution to losing telemetry, the New Horizons team waited. It began running through possible scenarios for why the spacecraft may have gone silent, noting for each scenario the intervals at which they would expect to hear back from it. If New Horizons had gone silent because of two simultaneous demands on the processor, for example, then the craft would reconnect within a certain window of time. If the scenario was more dire, the team would hear back within a different timeframe. So they waited. Bowman said, "In the case of New Horizons, waiting gave us insight about what was going on and time to plan how we could best resolve the issue."

The key hurdle at this particular moment was that the team needed to send the whole program up to the spacecraft that would tell it what to do upon approaching the planet. "Didn't we have this on our contingency list?" asked Sterne. Hersman replied that they did, but that the contingency plan would mean a totally different flyby for the spacecraft. Sterne, who had been working on this mission since he conceived of it in 1989, had a look of panic pour over his face. "Do we have to do that?" he asked. At this point, they were unsure, as they waited for the spacecraft to respond.

Here, on the Fourth of July, they had three days before they were supposed to begin capturing data about the planet. Forty-eight hours later, the spacecraft was back on track. With four hours to spare, the team had been able to run the sequence of commands in their ideal scenario for the closest approach to Pluto. In the team's wait time to hear responses from New Horizons, the scientists discovered that two systems had been using the same area of the memory at the same time, and the spacecraft had gone into a safe mode until it was told to come back online. All of this was determined at the slow pace of four-and-a-half-hour commands to New Horizons and four-and-a-half-hour responses.

The distance between explorers and the people to whom they are sending back messages creates a delay that serves as an "enabling constraint." Orson Welles famously noted, "The absence of limitations is the enemy of art." For Welles, constraints were enabling and were required for the imagination to do its work. Without constraints, art and innovation are paralyzed. Within the window of constraint, the true boundlessness of our explorations can be located. Finding innovative ways of working within the constraints of our current technologies helps us utilize them in new and unexpected ways. Constraints allow us to work within a narrow set of parameters to reconfigure expectations about how a tool is used to enable discovery.

Spacecraft currently use radio waves to communicate with Earth. While these waves can travel at the speed of light, they are weakened the farther they have to travel. The Jet Propulsion Laboratory in Pasadena, California, has been working on the next generation of communication technology for spacecraft: lasers.[13] A laser could produce a much higher bitrate than a radio wave, and critical data would be accessible much faster. Bowman explained that there is always a fear that something unknown will destroy the spacecraft, so after the Pluto flyby, everyone was on edge until the most critical data was downlinked. If the bitrate were increased significantly, that wait time would be reduced. The speed could be increased by a factor of one hundred. Just as our networked society has moved from dial-up internet connections to broadband, space programs will probably innovate their connections to space.

There is seemingly little advantage to an imposed constraint of waiting for messages from spacecraft. The advantages of quick communication can't be denied. But the wait times introduced by a long expedition like New Horizons do allow mission scientists and technical leads to create new protocols and adapt on the fly.

Even if laser communications could replace radio waves, it would take some time for NASA to implement them. In fact, nearly all the spacecraft produced for NASA, including the New Horizons mission, are built on old technologies.[14] These probes must use "space hardened" materials that can withstand the radiation of onboard power sources, such as plutonium, as well as the treacheries of space itself. For example, the decade-old hard drive designs used to store images and data from the mission had a capacity of only 16 gigabytes. As Hersman described the solid-state hard drives on this multimillion-dollar spacecraft, I marveled that I now carry a much larger drive on my keychain. The requirements for these spacecraft mean that they must employ technologies that are already far outdated compared with what is used in everyday life by the common consumer.

Similarly, the computers used to run the software on the ground for the mission were borrowed from a previous mission. These machines were so out of date that Bowman had to shop on eBay to find replacement parts to get the machines working. As systems have gone obsolete, JPL no longer uses the software, but Bowman told me that the people on her team continue to use software built by JPL in the 1990s, because they are familiar with it. She said, "Instead of upgrading to the next thing we decided that it was working just fine for us and we would stay on the platform." They have developed so much over such a long period of time with the old software that they don't want to switch to a newer system. They must adapt to using these outdated systems for the latest scientific work.

Working within these constraints may seem limiting. However, building tools with specific constraints—from outdated technologies and low bitrate radio antennas—can enlighten us. For example, as scientists started to explore what they could learn from the wait times while communicating with deep space probes, they discovered that the time lag was extraordinarily useful information. Wait times, they realized, constitute an essential component for locating a probe in space, calculating its trajectory, and accurately locating a target like Pluto in space. There is no GPS for spacecraft (they aren't on the globe, after all), so scientists had to find a way to locate the spacecraft in the vast expanse. Before 1960, the location of planets and objects in deep space was established through astronomical observation, placing an object like Pluto against a background of stars to determine its position.[15] In 1961, an experiment at the Goldstone Deep Space Communications Complex in California used radar to more accurately define an "astronomical unit" and help measure distances in space much more accurately.[16] NASA used this new data as part of creating the trajectories for missions in the following years. Using the data from radio signals across a wide range of missions over the decades, the Deep Space Network maintained an ongoing database that helped further refine the definition of an astronomical unit— a kind of longitudinal study of space distances that now allows missions like New Horizons to create accurate flight trajectories.

The Deep Space Network continued to find inventive ways of using the time lag of radio waves to locate objects in space, ultimately finding that certain ways of waiting for a downlink signal from the spacecraft were less accurate than others. It turned to using the antennas from multiple locations, such as Goldstone in California and the antennas in Canberra, Australia, or Madrid, Spain, to time how long the signal took to hit these different locations on Earth. The time it takes to receive these signals from the spacecraft works as a way to locate the probes as they are journeying to their destination. Latency—or the different time lag of receiving radio signals on different locations of Earth—is the key way that deep space objects are located as they journey through space. This discovery was made possible during the wait times for communicating with these craft alongside the decades of data gathered from each space mission. Without the constraint of waiting, the notion of using time as a locating feature wouldn't have been possible.

The dominant technology of an era becomes that era's lens into the inner workings of the human body. For the Greeks, the water clock led to the concept of "humors," bodily fluids that dictated how people felt and behaved. As societies developed mechanical clocks, people began imagining the inner workings of the body as clockwork or machinery. As electricity became pervasive, the metaphor of electricity was used to understand how our nervous systems worked. In our own era, we think of the body through the lens of the computer, as we "process information" or need to "reboot." Another result of this incorporation of a technology into the body is that our minds expand through this tool to imagine new ways of doing familiar tasks. The result is a defamiliarization of the familiar; things that we looked at every day are seen through new eyes. It is through these mental expansions as we interact with technologies that we are able to innovate.

What about the converse: How do our brains adapt when our familiar tools are no longer viable or usable? How do we adapt to constraint? Constraints work in similar ways to new tools in that they foster "defamiliarization" with existing tools. By interacting with a tool that imposes constraints, we are given a bounding frame within which we are able to imagine new uses. These new uses become new tools in and of themselves, as in the case of using radio waves from spacecraft as a location tool or to analyze a planet's atmosphere. Radio, which was meant to send communications data, is now repurposed to become something altogether new.

For scientific explorations like New Horizons (or the longitudinal study of Harvard students), wait times are an enabling constraint that allows scientists and engineers to imagine new ways of working within those boundaries. Constraints are enabling and encourage us to think expansively. Cognitive psychologists have dubbed this function of the imagination the "default network" of neurons in the brain, because it typically fires when our brains are not occupied by other tasks (that is, when we have to wait). When this web of neurons is activated through the time lag introduced by wait times (including the boredom or daydreaming during the in-between times of active tasks), the psychologist Jonathan W. Schooler proposes, "our imaginings may allow us to stumble on ideas and associations that we may never find if we strive to seek them."[17] Once these new associations are created out of enabling constraints, new knowledge can be discovered. As Manoush Zomorodi writes in *Bored and Brilliant*, "The default mode is not surprisingly called the 'imagination network.' Being bored gives us the space to ask 'What if?' That's an essential question regarding not only any creative endeavor but also our emotional health and personal growth."[18]

After these ideas are created and knowledge is formed, the human brain also requires wait times in order to process those ideas and turn them into lasting memories that can be later accessed. In 1966, the neurobiologist James L. McGaugh reported the foundational research that has shaped how we understand the pathways knowledge takes from short-term and working memories to long-term memories accessible years after learning. For McGaugh, memories "are not laid down in any lasting way either during or immediately after the experience."[19] Instead, "memory consolidation," which transforms a memory into something more durable and lasting in the brain, occurs over long periods of time. What McGaugh discovered in the mid-1960s was that as the time increased between learning something and being tested, retention increased. He wrote, "It has long been known (and ignored) that, within limits, learning is facilitated by increasing the interval between repeated trials."[20] As demonstrated in his studies, memory consolidation takes time, and the ability to form knowledge that stays in the brain for the long term requires waiting. Accessing knowledge later for use in building on that knowledge requires that the memory be given the time to move from the short-term memory regions of the brain to the working memory regions to the long-term memory regions. As educational consultant David A. Sousa writes, short-term (or "immediate") memory is similar to a clipboard on which our brain temporarily stores ideas until it decides what to do with them. This is typically an unconscious mental process, where the brain decides rapidly whether an idea or memory should be passed along to the working memory or simply discarded. Once passed along to working memory, these memories and ideas often come under the purview of the conscious mind. Working memory is like a large table where items are laid out and dealt with; adults often handle these memories within ten or twenty minutes. Sometimes, we have unresolved questions or thoughts that will stay in the working memory for hours or days if they are not dealt with (that is, if they aren't discarded or passed along to long-term memory).[21]

Once ideas or memories are passed along to the long-term memory portion of the brain—typically because the ideas "make sense" or "have deep meaning" for the person, as Sousa notes—they can be accessed again later and combined with other ideas to build on knowledge. However, storing information for the brain takes time and the process improves as we increase the gaps between handling this information. This is why, for example, most researchers advocate that students should not cram for an exam; instead, by spacing out what they learn—by waiting between study sessions—they are able to retain more information. Cramming keeps information in the working memory, but it is quickly discarded unless the time is given to allow it to be encoded through the hippocampus region of the brain, which "sends it to one or more long-term storage areas," as Sousa notes. "The encoding process takes time. . . . While learners may *seem* to have acquired the new information or skill in a lesson, there is no guarantee that storage will be permanent after the lesson."[22] Retention, he notes, will probably occur if the student can recall the knowledge after the time when the "greatest loss of newly acquired information or a skill occurs"—within the first eighteen to twenty-four hours after acquisition.

Building on this work, NYU researchers in the Center for Neural Science have published a 2017 study that supports the idea that memory takes time and that the very structure of how we make knowledge into long-term memories is, at its core, a time-based problem. In this work, appropriately titled "Memory Takes Time," the consolidation of memories takes so long because it requires RNA and/or protein synthesis, a process that is "temporally restricted." So "in order for lasting changes in the synaptic state to persist," time must be given to these processes.[23]

In other words, a fact we quickly learn through a search engine will often not make its way from working memory to long-term memory unless it is given the time to do so. As we engage with knowledge through repetition of practice, and as ideas make sense and have meaning in our lives, we must then give them time to settle into our long-term memories.

Everything we know is built on some form of waiting. From waiting for explorers to send back word about the unknown, to the speculations that allow us to innovate on existing knowledge, to transforming that knowledge into long-term memories that can be accessed and built upon, wait times are essential for the things we know and for creating new knowledge.

We similarly engage in this kind of knowledge building with one another, as we send messages across distances to keep in touch. We learn about one another through the messages sent and build personal relationships through the wait times often imposed on our lives as social beings.

5 A DELAYED CROSSING

This past winter, I stood on the Fredericksburg side of the Rappahannock River in Virginia, next to the railroad bridge leading into town. In the early 1860s, this bridge was one of the first casualties of the American Civil War, blown up by the Confederate Army to impede easy entry into the city by federal forces. The same bridge, now rebuilt, carries Amtrak trains into the Fredericksburg station and long freight trains north to Washington and Baltimore. A restaurant sits to one side of the bridge, its patio furniture stacked until warmer weather arrives. I walked through the restaurant's parking lot, past a full Dumpster, to the slope leading down to the river. Standing under one of the supporting arches of the bridge, I looked up and noticed wires and cables attached to the brick-and-concrete structure. The fiber optic cables that bring the internet into cities and towns across the United States are able to cross rivers and bays by following the routes created by Civil War–era bridges. The fiber optic cables attached to the side of this bridge in Fredericksburg are inside a large metal pipe that runs the length of the bridge. Approaching the shore, the pipe bends downward and the cables go underground right where the bridge meets land. It's easy to spot fiber optic cables like these along the historic bridges in many parts of the United States, but they typically go unnoticed by the people who pass them each day. Our ideas of how the internet gets into our homes rarely include an image of wires clinging to Civil War–era bridges.[1]

I made the two-hour drive from my home in the Washington, DC, area to Fredericksburg to visit a handful of sites that I hoped would tell me more about how our experiences of delay have changed. As the messages we send each other are delivered faster and more efficiently, how has this shift changed our sense of being connected with one another? How was communication different in December 1862, as soldiers attempted to cross the Rappahannock River during the Battle of Fredericksburg? Soldiers sent letters back home at an unprecedented pace during the Civil War, but they would often wait weeks or months for a response.

Looking at the cables on this bridge, I think about all the data coming into and leaving Fredericksburg: the emails, comments, videos, messages, pictures, and other media we exchange at a seemingly instantaneous pace. Just before parts of this bridge

Figure 16. A metal pipe carrying fiber optic cables across the Rappahannock River in Fredericksburg, Virginia, attached to the Civil War–era railroad bridge. Image © 2016 Jason Farman.

were destroyed by Confederate forces, the Richmond, Fredericksburg, and Potomac Railroad used this bridge to send mail across the state, along with passengers and supplies. As I explored Fredericksburg, comparing these two different eras of communication, I began to notice striking similarities between our data-driven messages and the letters shuttled across the Rappahannock.

At this railroad bridge in Fredericksburg, three factors come together: human connection, time, and geography. Finding ways for these three factors to intersect has been the challenge for human communication throughout history. Bridges and messages are tools that connect people across geography. They allow us to be intimate despite the miles that separate us. They allow us the ability to imagine the possibility of linking up, representing the hope we have in eventually reaching each other. They connect us across vast spans. Bridges and the messages they convey, whether they are emails sent across fiber optic cables or letters delivered on railroad cars, symbolize the core concerns that have led me around the world in search of the ways that waiting has become an essential part of the messages we send.

The Battle of Fredericksburg was a disastrous defeat for federal troops because key messages arrived late, which in turn led to the late arrival of supplies needed to cross the Rappahannock. These delays allowed word to get to Robert E. Lee that the Union troops were about to cross into Fredericksburg. He fortified the city in time for the Union's "surprise" attack, which lasted four days and left thousands of soldiers dead, many falling just north of this railroad bridge. There, standing on portions of a makeshift pontoon bridge used to usher in troops and artillery, army engineers attempted to complete the bridge under heavy fire from Confederate sharpshooters.

The bridge builders and engineers during the Civil War created improvisational connections to enable troop movement. Bridges were an interface that allowed ad hoc networks to be built, connecting distant areas and their people. Similar to the ways that network engineers created "quick and dirty" ways of laying fiber optic cables, as described by Randolph Stark in chapter 2, pontoon bridges allowed for inventive connections that could happen on the fly. Movement and speed relied on these kinds of connections to be rapidly deployed and were designed to keep troops from waiting. Yet issues of capacity and of delayed messages meant that these quick deployments ultimately produced long wait times.

As I read about soldiers stationed here in November and December 1862, writing home about their delayed crossing of the Rappahannock, one soldier symbolized waiting for me in a profound way: Private Joseph Coryell of the 24th Michigan Volunteer Infantry Regiment.

Just over a month before the Battle of Fredericksburg, Coryell saw a pontoon bridge for the first time. This makeshift bridge constructed out of wood planks draped across floating pontoon boats spanned the Potomac River in Berlin (now Brunswick), Maryland. The bridge, and the long line to cross it, was a spectacle like nothing Coryell had seen before. He was a farmer from Clinton County, Michigan, who had joined the 24th Michigan Regiment two and a half months earlier. As the 24th Michigan, now a part of the famous Iron Brigade, approached the Potomac, the line to get across stretched back four miles long. Several days earlier, the soldiers in his regiment were ordered to march at an incredible pace—packing light and sleeping outdoors with no tents—in order to catch up with the rest of the Army of the Potomac, commanded by Major General George McClellan. The 24th Michigan had rushed to get to Berlin, moving hurriedly past the horrific remains of the Battle of Antietam, only to get stuck in traffic approaching the bridge. An "immense body of men, animals, wagons, and cattle crossed in one continual stream," all day and night, one of the army engineers wrote about the Berlin crossing.[2] The 24th Michigan had to wait almost four days for its turn to cross the Potomac.

Coryell's description of the miles-long line of Union soldiers caught in this traffic jam illustrates the first half of an old adage: "War is long periods of boredom punctuated by moments of terror." Coryell wrote home saying, "The soldier's life is a lazy one," despite having written in another letter that he had marched more than two hundred miles since being deployed. Waiting, instead of marching or moving or fighting, characterized his time as a

Figure 17. A pontoon bridge across the Potomac River north of Fredericksburg, Virginia, May 1864. Image courtesy of the Library of Congress.

soldier. He waited so much as a private in the army that he had gained an enormous amount of weight, eating while bored. For these troops in the Army of the Potomac—in contrast to their Confederate counterparts—food was plentiful, and so was time.

However, for Coryell and the other troops who were waiting days to cross a bridge in Maryland, the traffic delay was not a sign of disorganization or idle waste of time; instead, it was a sign of how big the Army of the Potomac actually was. Waiting meant that the troops were numerous. It was a sign of their strength. For many, such a sight was a welcome relief in contrast to the bloodshed at Antietam and Bull Run. The weather was sunny, spirits were high, and everything seemed to be pointing toward a "speedy and favorable termination of the War."[3]

For Coryell, waiting and delays would shape his entire experience in combat during the Civil War. The delays he endured in

Berlin, Maryland, before crossing the Potomac would be played out again and again throughout the region during his time in the Iron Brigade. Delays in letters that kept him connected with his wife back in Michigan and delays crossing bridges were the most notable. But his day-to-day life as a soldier was also shaped by other experiences of waiting, such as delays in marching orders, when soldiers were told to pack their supplies, but then wait for instructions with no clue about where they were headed next. Coryell would wait for his paycheck from "Uncle Samuel" (as he called the U.S. government), which would sometimes be six months overdue. He would wait out winter weather in a log "shanty" he built in northern Virginia, among a small village of similar log dwellings thrown together by the soldiers. He would wait for horses to pull wagons out of deep muddy ruts on the roads, as they marched through days of rain. In the end, delays like this ultimately led to his death along the Rappahannock River, just over four months after the Battle of Fredericksburg.

Coryell's actions didn't shape the course of events in the Civil War, nor did he make choices that were any more heroic than his fellow infantrymen. He may never even have shot his weapon in combat. But his descriptions of the war and his internal experiences handling the conflict as a thirty-two-year-old farmer from Michigan transformed the ways I understand the impact of delay on the course of human events. Take, for example, the delays of the Post Office Department in conveying letters between Coryell and his wife, Sarah, at their farm in DeWitt, Michigan, just outside Lansing. As Coryell waited for word, the entirety of the war waited in similar ways, sending messages across the United States at a scale that was unprecedented. Waiting and delay—for Coryell, for soldiers and families on both sides of the conflict, for enslaved African Americans waiting for emancipation, even for those in the highest seats of power at the time—were experiences that were woven into the very fabric of life.

I first came across Coryell's letters in a library in Orange County, California. I walked into the library on an intense summer day in the middle of California's worst drought in memory, passing a

Newsweek headline that read, "NASA: California Has One Year of Water Left." I went upstairs and sat down at a table in the Special Collections division of Leatherby Library at Chapman University. Here, in a dry, air-conditioned room, I was handed an unsorted box of Civil War letters. In the months before my arrival, the library was given a massive corpus of war letters by a Washington, DC, historian who had been collecting them for decades. This historian had written to the Dear Abby advice column about a fire that consumed his family's home.[4] Inside the house were all the family's letters, including correspondence from wartime. He concluded by urging readers to preserve their letters by sending them to his newly formed organization. The wartime letters began to pour in. Continuing the plea for families to send in their letters through the organization's website, he soon amassed around ninety-eight thousand letters from various American wars, some dating back to 1847 and the Mexican-American War.

When I first visited the library, the letters had just arrived from Washington and were still in the crates and bins in which the historian had sent them to the university. Opening one of these bins revealed a trove of unsorted letters detailing fascinating personal stories of soldiers from nearly every conflict in which the United States had been engaged. Until now, only the soldiers' families and, for some, the Washington historian had ever read these letters. None were digitized or sorted. Some were even still in their envelopes, postmarked more than a hundred and fifty years earlier.

I read through hundreds of the letters looking for soldiers' accounts of waiting for messages from loved ones. I was tracing the time it took for messages to be delivered at different times in American history and how people's expectations shifted as technologies—especially in transportation such as railroads, paved roads, and eventually air mail toward the end of World War I— shortened the amount of time for deliveries. For example, during the first three months that Edwin Biles was a soldier in Puebla, Mexico, in 1847, he hadn't received a single letter from home. He sent long letters to his family and friends back in Philadelphia, but heard nothing back. Finally, a few days after the Fourth of

July, he received a letter from his father dated three months earlier. In his reply to his father, Biles wrote, "I received your letter dated 21st March last evening, so you see that it has been more than three months on the roads." For these months that Biles manned a twenty-four-pound howitzer cannon during several battles in the Mexican-American War, he tried to figure out why there was silence from his loved ones. Now, knowing that the delay of his father's letter was due to the slow crawl of the postal workers on unpaved roads across two countries, Biles was relieved. He wrote, "I had been expecting to hear from you for some time, and could not imagine why I did not receive your letter. I am very glad indeed to hear that I am not forgotten."

Joseph Coryell experienced something very similar to Edwin Biles. Coryell was a prolific writer; he wrote letters to his wife with a frequency that outpaced most of the other Civil War soldiers I came across in the archives. Yet during the first part of his service, he never received a response. In his first letter home to DeWitt, dated September 1, 1862, he wrote that he had left Detroit in such a hurry that he was unable to make it to the post office to send a letter home in time before they departed. Now, stationed just outside of Alexandria, Virginia, he was finally getting a chance to send her the letter he had written. Each evening, as the troops were stationed along the Potomac or Rappahannock River, a mail boat would pull into camp and the postmaster would call out, "Hurry with your letters!" This happened each evening around 4:00 P.M. The boat would leave and, as Coryell describes it, "then [the letters] are taken to the brigade headquarters, from there to division headquarters, from there down to the landing and the same boat that brings in our mail takes our letters out the same night." The boat would return four hours later as the postmaster returned with incoming mail. Coryell wrote, "You do not know with what anxiety I listen every night to hear my name called when the mail is being distributed." However, during the first month of his enlistment, he didn't receive a single letter from home. He couldn't quite be sure how long letters were taking en route to Michigan until other soldiers began receiving theirs from Clinton and Wayne Counties, home to most of the soldiers in the 24th Michigan. As the letters came in, the soldiers would look at the postmark and note that, on average, letters were taking ten to twelve days to arrive.

The pace of life for these soldiers revolved around the various ways they were connected with others in distant locations. Mail was a way to divide and order the day. Letters arriving were markers of a specific time of day and were also objects that connected the soldiers with their former lives back home. The letters were bridges to the life a soldier knew outside of the horrors of war. As historian David Henkin puts it, "More than in any previous war, using the postal network was a major part of the Civil War experience, both for those who fought and for those who followed the conflict through correspondence with men (and, to a much lesser extent, women) at the front."[5]

When Coryell left DeWitt to join up with the rest of the 24th Infantry, he sat down to write a quick departing note to his wife. He was in Detroit, which was more bustling than usual, with huge crowds lining the sidewalks ready for the farewell parade seeing the 101 men leaving for the heat of the battle below the Mason-Dixon Line. The noise was immense, loud enough to distract someone from focusing on writing a letter. This was especially true for Coryell, who had grown up on a farm just below the Finger Lakes region of New York and who had moved to Michigan in his twenties to start a farm of his own. Farm life contrasted his experience in Detroit, as he sat on his rolled-up blanket, with a scrap of wood as an improvised desk on which to write. As he wrote the letter, the parade started and the men had to gather to board two steamboats across Lake Erie to Cleveland. He put the letter into his pack and joined the other men so as not to miss his obligation in the march down Woodward Avenue.

The brigade left Detroit on August 29, 1862, with many thousands of onlookers cheering them on. Several states away, the second day of fighting was under way during the Second Battle of Bull Run or Manassas in northern Virginia, a crushing defeat for the Union. The Michigan volunteers had enlisted to serve as reinforcements for the soldiers in the Army of the Potomac, who were this day fighting Stonewall Jackson at Bull Run. The following day, the Union Army would retreat.

The path the 24th Michigan would take to join the Army of the Potomac in northern Virginia was nearly the exact route that the letters back home would travel. In camp, wherever he was stationed, Coryell would hand the postmaster the sealed envelope (often with no stamp affixed, leaving the postage to be paid by his wife back home, since stamps were in high demand in camp and difficult to come across) and visualize the journey the letter would take up the river into Washington, along the B&O Railroad through Baltimore and into Harrisburg, Pennsylvania. From Harrisburg, it would follow the rivers and roads to Pittsburgh and up to Cleveland. There, the boat would ferry the mail across Like Erie to Detroit and then out to the neighborhood post offices. There were no street addresses on these letters. Coryell simply addressed his letters: Mrs. Sarah Coryell, DeWitt, Clinton County, Mich. In 1863, one postal worker in Cleveland was so moved by seeing lines of women lined up in the cold hoping for a letter from a husband, son, or brother that he started the practice of delivering letters to homes with a knock on the door.[6] This was the first time that home delivery was employed in the United States and led to Congress approving free home delivery of the mail that same year. The rise of home delivery of the mail was, in large part, a response to the uncertainties of correspondence during the Civil War. Most of Coryell's letters were left at the DeWitt Post Office to be picked up by his wife or a relative.

His first letters were not collected by his wife, but instead were picked up by Coryell's younger sister, Mary Jane, who had married a local farmer, Andrew Dunlap. Their farms bordered each other, and initially, all of Joseph's letters were intercepted by his sister. Mary Jane opened the first letter and, noticing that Joseph had written only on half of the paper, filled the rest with her own note to her sister-in-law. As she wrote the note, she was sitting across the table from Ruth, the wife of another soldier in the 24th Michigan, James Hubbard. James was also a local farmer and was Joseph's closest friend, the two having enlisted on the same day. Joseph's sister wrote in the note, "Ruth is here and we have been writing a long letter to Jim and Joe. We do not know [whether] it will ever reach them but we thought they would feel so uneasy about the money [they had sent in their letters] we had better try it anyway. Joe did not give any directions [about where to send our letters] but Jim said direct to Washington, DC, 24 Mich Infantry, Co. F."

People back home were often unsure of where letters should be sent to soldiers who were out in fields seemingly in the middle of nowhere. They were especially concerned that the letters would never reach their loved ones. Despite not hearing back from Sarah (perhaps due to her lack of knowledge about where she should send letters), Joseph wrote to her every other day during the first month of his deployment. A month after he enlisted in the army, he wrote a letter home saying, "My dear beloved wife. I have heard nothing from home since I left there. . . . I wish you would write very often and then maybe I will hear from you once in a while." He wrote many letters in the hopes that at least one of them would actually make it to her hands; here, he recommends that she do the same.

On September 20, 1862, the postmaster finally called out Joseph Coryell's name. He received his first letter from Sarah thirty-eight days after he had enlisted and twenty-two days after his regiment had left Detroit for the war. He was overjoyed. However, Sarah's letter spoke only of her sorrow at having her husband in the war. She had already lost hope in ever seeing him again, especially in the wake of the massive number of men killed in the recent conflict at Bull Run. She feared that her three-month-old son would never know his father. Coryell closes his next letter by asking her to put her trust in God: "Pray to him without ceasing that he will preserve us and bring us home in safety. Pray to him that he will bring this wicked war to a speedy close, that he will put it into the hearts of the rulers of this nation to do all they can to bring it to a close and that he will root out rebellion from our land and that peace may soon be restored."

For Coryell and many soldiers like him, these weren't simply religious clichés tossed around to justify war. His decision to fight in the war arose from his religious views as an abolitionist. Around the time that Coryell had moved to Michigan to start a farm, with the hope of starting a family of his own, two events had altered the course of his life. The first was the report spreading through newspapers and correspondence in the North about the breakup of families of slaves. One quarter of every slave family was broken apart, according to the reports, each person being sold individually. James McPherson, the well-known Civil War historian, noted, "The sale of young children apart from parents, while not the normal pattern, also occurred with alarming frequency. This breakup of families was the largest chink in the armor of slavery's defenders. Abolitionists thrust their sword through the chink."[7]

Second, around the time Coryell moved to Michigan, the United States passed the Fugitive Slave Act of 1850. This new law allowed southerners to cross state lines, even into a free state, to recapture their "escaped property." As a result, many free black men and women were essentially kidnapped from the North and brought into slavery, since the burden of proof was on African Americans to prove their free status. Many religious believers in the North were horrified by the treatment of God's creation, scandalized that people were being put into bondage against their will. Many quoted Galatians 5:1, "Stand fast therefore in the liberty wherewith Christ hath made us free, and be not entangled again with the yoke of bondage." Slavery and the dissolution of the family, especially separation of the young, were direct affronts to the biblical concept of liberty. For Coryell and others like him, the United States had become a nation with practices that were in direct contrast to his own moral compass. He and many he enlisted with believed that America's practice of slavery was a disgrace to God.

So on August 13, 1862, he left his wife and two-month-old son, Johnny, to enlist in the war effort. Standing with his friend Jim Hubbard and almost every farmer, student, blacksmith, and carpenter in Clinton County, Michigan—as all throughout the northern states—his decision was to delay the life he had established in hope of creating something better.

The result was hundreds of thousands of men moving about the country, traveling on railroads, by boat, and on foot to new parts of the nation, mostly to places they had never seen before. The 24th Michigan crossed over seemingly arbitrary state lines of

Ohio and Pennsylvania (solidly northern states) and into Maryland (one of the most uncertain states, home to a mix of northern and southern supporters), finally entering Washington, DC, and northern Virginia. "Old Virginia," as Coryell and his fellow soldiers called it, was clearly on the opposite side of the line of conflict. The distances the 24th traveled could be bridged only through the messages the men sent back home in their letters. And with ample time on their hands as they awaited marching orders or waited in line to cross a bridge, they wrote countless such letters.

The arrival of these letters and the crossing of bridges have much in common. Each is a symbol of connection and delay in the Civil War. Crossing bridges is the only way for movement to happen in much of Virginia, Maryland, and the Washington area. But these pathways presented constant challenges for the armies. They confronted a challenge that I discussed in chapter 3 of this book, the problem of "bandwidth"—the volume of people, wagons, artillery, and supplies you can get across the bridge depends on the load it can accommodate. That load will cause a delay if it exceeds the bridge's capacity. Similarly, letter delivery had a specific bandwidth at the time, often confronting the exact same challenges that the troops were facing: destroyed railroad bridges and muddy, impassible roads.

Delay was therefore an organizing fact of life during the Civil War. Coryell's wait to cross the pontoon bridge at the Potomac in Berlin, Maryland, was a metaphor for the life of a soldier in the army. They were not only waiting to cross over into Virginia; they were also waiting for the war to end so they could get back home to their normal lives. Their lives were on hold; by enlisting in the army, the soldiers were delaying their everyday lives for the causes they were fighting for. The entire experience was one of delay.

Joseph Coryell crossed the pontoon bridge into Virginia on October 30, 1862. Days later, General McClellan would be relieved of his command of the Army of the Potomac in part because he waited too long to pursue the retreating Confederate forces after the Battle of Antietam, an action that would have enabled him—at least as President Abraham Lincoln saw it—to crush the Confederate

Army. McClellan's delay probably extended the war by years, Lincoln argued, and the president replaced him with Ambrose Burnside in early November 1862. Burnside took McClellan's delay as a lesson and would not be guilty of the same inaction.

The day after he took command, Burnside mailed a request to have the pontoon bridge moved from Berlin to Falmouth, just across the Rappahannock River from Fredericksburg. Before he arrived, he received reports that the stone bridge from Falmouth, Virginia, into Fredericksburg had been destroyed. In order to get across the Rappahannock, he would have to build pontoon bridges, a tactic used extensively during the Civil War. Fording the river, either with men on foot wading across or with wagons being pulled across, would be a slow and dangerous prospect due to the rocky floor of the river at its shallowest location. The men would have to walk cautiously and some wagons would probably tip over on the uneven riverbed. For Burnside, fording the river wasn't an option. So before arriving in Falmouth, he sent his request to his superiors for the pontoon bridge supplies and the engineers to be delivered from Washington. He sent this order via mail as he moved the troops through Warrenton, Virginia, and then onto the shores just across from the city, ready to make his surprise attack across the four hundred–foot span of river. He would then move south from Fredericksburg, he determined, ultimately conquering Richmond and ending the war. Delaying action had been McClellan's undoing, and Burnside worked at a fever pitch to make sure the same would not be said of him. However, looking over all the many histories that have been written about the Civil War, perhaps the greatest example of disastrous delay was Burnside's long wait for the pontoon bridges at Fredericksburg.

Before the 50th New York Volunteer Engineers regiment received instructions to move the pontoon bridge from Berlin to Falmouth via Washington, it was almost forty miles from the nearest federal troops. The weather was gorgeous and the men relaxed on the shores of the Potomac, awaiting their orders. They had until this time been throwing bridges across rivers at a nonstop pace, most recently in Harpers Ferry next to the demolished railroad bridge that had paved the way for John Brown's failed raid on the federal armory there. Colonel Wesley Brainerd wrote of this time on the Potomac after the army was well on its way to Warrington and Falmouth: "Day after day we waited for orders and wondered why they did not come. Gradually, we accumulated our materials and put things in readiness for instant departure, feeling that our orders *must* come before long. Thus the beautiful Autumn days wore along, but no news from the Army. We were completely isolated, perhaps forgotten."[8]

Nearly a week after Burnside issued the orders, a courier brought in a mailed letter (not a quickly dispatched telegram) to the 50th New York Regiment, instructing it to move to Washington with the entire bridge train. By the time the message arrived, the unit was already supposed to be in Washington, according to Burnside's orders. Brainerd explains that in the confusion of shifting command from McClellan to Burnside, "the message was laid aside or held in abeyance until the designs of the new commander should take shape. Hence this *fatal* delay which cost us the loss of a terrible battle and plunged the whole country into depths of despondency and gloom from which it did not recover until the end of the final termination of the war."[9]

The 50th New York moved quickly to Washington, where it was then ordered to transport the pontoon train to Falmouth. The regiment left Washington on November 19, two weeks after Burnside had ordered the bridges moved. Half of the bridge supplies went to the site via the Potomac to Aquia Creek and then on to Falmouth. The other half went by land, struggling along muddy roads and impassable pathways. As the unit journeyed, it received daily letters saying, "Hurry up! We are in desperate need for the pontoon bridge."

Burnside, along with the entire Army of the Potomac, had been stationed across the four hundred–foot expanse of river from Fredericksburg for nearly three weeks in anticipation of his attack. His plans initially looked promising. In taking Fredericksburg by surprise, the Union would face little resistance from the five hundred to one thousand soldiers thought to be stationed in the town. Robert E. Lee and the Confederate Army were dispersed and unaware of the Union's location. Burnside initially moved quickly and was going to push the Union forward and end the war.

Instead, Burnside was forced to wait.

According to one early report to Burnside, the pontoon bridge supplies had left Washington on November 7, the day after he requested them. The fifty-mile journey from Washington to Burnside's location should have taken about five days. Weeks later, the pontoon bridge still had not arrived. The general's plans were so significantly delayed that by the time the 50th New York Engineer Regiment arrived with the two loads of pontoons and planks on November 25, Lee's troops had arrived in Fredericksburg and fortified the city with more than seventy thousand men, with new troops arriving by the hour, in anticipation of an attack from Union forces. The engineer regiment would start to build the pontoon bridge at 1:00 A.M. on December 11, exactly one month after Burnside had expected the pontoon bridge to arrive.

On that day, the civil engineers in the 50th New York Regiment were halfway out onto the Rappahannock River in Virginia, standing on an incomplete bridge made of planks of wood strung across floating pontoon boats. Bullets were flying past them, spattering the water and splintering the wood on the makeshift bridge. The shots were coming from Confederate soldiers hidden in the thick fog that covered the opposite bank of the river. The sun had not yet risen, but the engineers were working at a rapid, nervous pace to get the bridge built, allowing the Union Army to make its "surprise" crossing en masse into the town of Fredericksburg. As Captain Augustus Perkins of the 50th New York guided one of these pontoon boats into position, he was shot in the neck by a sharpshooter stationed across the river.[10] He was the only officer of the 50th New York killed and may have been the first casualty of the Battle of Fredericksburg. After Perkins was shot, other soldiers around him who were attempting to build this floating wooden bridge were killed with little opportunity to fight back.

Having begun the bridge building as the church bell across the river rang out 1 o'clock in the morning Colonel Brainerd and his 50th New York Engineer Regiment had completed half of the four hundred–foot expanse by 3:00 A.M., seemingly unnoticed by enemy troops. An hour later, the bullets from the Confederate troops "rained upon [the] bridge." Brainerd continued, "They went whizzing and spitting by and around me, pattering on the bridge, splashing into the water and thugging through the boats. Where were my men? They did not require any command to fall back in good order." The engineers scattered, fleeing the bridge and seeking shelter out of the range of the guns. Brainerd describes the scene by saying, "Every one started for the shore end of the bridge. Some fell into the boats, dead. Some fell into the stream and some onto the bridge, dead. Some wounded, crawled along on their hands and knees and in a few moments all of us were off the bridge, all except the dead. The storm of lead continued."[11]

Burnside made a bold decision in response to the sharpshooters: He became the first American military general to order the bombardment of an American city.[12] The cannons opened fire on Fredericksburg, and the shots from the sharpshooters ceased. The 50th New York cautiously stepped back onto the wood planks of the pontoon bridge and resumed its building. Once again, the shots started flying toward the men. Again, the engineers fled. Again, the cannons from the Union artillery responded.

As the men of the 50th New York scattered from their hammers, fleeing from the gunfire of the sharpshooters of the 13th Mississippi (known as Barksdale's Mississippi Brigade), it seemed that the bridge would never be completed on that December morning. The artillery that was bombarding the town of Fredericksburg in response to the sharpshooters was stationed on the heights across the river from the entrenched Mississippi gunmen. The guns could not aim low enough to do much good in dislodging Barksdale's Brigade of fifteen hundred men. Artillery was redeployed closer to the shoreline of the river, again with little effect. Finally, the fog began to lift just after noon. In the light of the rapidly warming December day, the Union troops were able to see exactly where the sharpshooters were located. Two Union regiments—the 7th Michigan and the 9th Massachusetts—got into pontoon boats and began to row across in an attempt to root out the enemy forces keeping the 50th from completing the bridge. The pontoon boats were designed for supporting wood planks for a bridge and were ill equipped for rowing across a river. As the soldiers slowly made their way across the Rappahannock, some paddling with the butts of their rifles, they took on fire. Making the first beach landing under fire in U.S. military history, they ran up the shore, capturing about thirty Confederate soldiers, with the rest of the "rebs" fleeing into the streets of the town.[13] Finally, the 50th New York Regiment was free to finish building the pontoon bridge, allowing the Union forces to cross into Fredericksburg. Urban warfare commenced as the Union troops entered the streets, taking fire from gunmen hiding behind upper windows of the houses. The battle would continue for several days.

On the first day of the conflict, Joseph Coryell and the 24th Michigan were stationed in Falmouth, across the river from Fredericksburg, next to the cannons bombarding the town. Just before the engineers began building the pontoon bridge across the Rappahannock, Coryell wrote a short letter to Sarah on a lined sheet of paper with a red and blue printed image in the upper left corner: an American flag with text underneath reading "Protect It." In this short letter to his wife, he writes about sending money home and doesn't even reference the impending battle.

A few days later, while he was out on the battlefield, he received a letter from Sarah. He wrote her back immediately:

> Fredericksburg   Dec 15, 1862
> Dear Wife,
> I received your long letter this morning together with a letter from Andrew and Jane. We are on the battle field now and have been these three days. Three nights we have lain down in the line of battle. We are on the extreme left wing holding our position merely. Our loss in the 24th has been seven killed and fourteen wounded so far but we don't know how soon we will be attacked but it is the impression of some of the officers that the rebs are on the retreat. Jim and I are all right. So far there has been none of the Clinton [County] boys hurt as yet and God grant that we may not be. I think that it is a false alarm about the rebs retreating, but I hope they will retreat until there will not be a rebel in the United States in a fortnight. Your letter was a good one just such as I like to get. I like to have you tell me about every thing, how you are getting along and how Andrew's folks are getting along. Jim

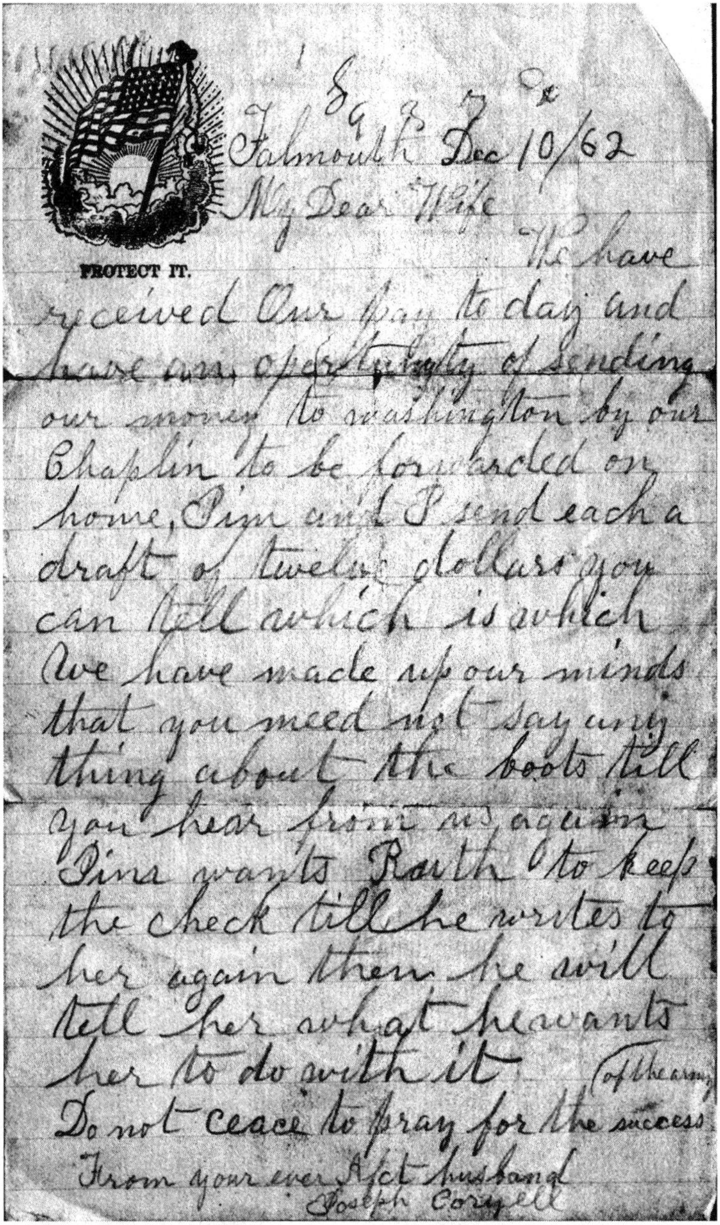

Falmouth Dec 10/62

My Dear Wife

PROTECT IT.

We have received Our pay to day and have an opportunity of sending our money to washington by our Chaplin to be forwarded on home, Jim and I send each a draft of twelve dollars you can tell which is which We have made up our minds that you need not say any thing about the boots till you hear from us again Jim wants Rath to keep the check till he writes to her again then he will tell her what he wants her to do with it (of the army) Do not ceace to pray for the success From your ever Afct husband Joseph Coryell

Figure 18. A letter from Joseph Coryell to his wife, Sarah, written on December 10, 1862. He wrote this letter on the day before the Battle of Fredericksburg, hunkered down in anticipation just across the Rappahannock River. Image courtesy of the Center for American War Letters Archives, Chapman University.

and I are well and hearty and we are a considerable fleshyer than we were when we left home not withstanding the hard fare. All the fight our regiment has seen as yet has been artillery fighting and our loss was by shells caused by supporting batteries. I must close for the mail is ready to leave. God bless you. From your affectionate husband,
Joseph Coryell

Hours after Coryell penned this letter, with twelve hundred Union soldiers now dead and "piled like logs" in mass graves in Fredericksburg, Burnside quietly withdrew his troops back across the Rappahannock.

Four days later, in his testimony before a congressional inquiry about the defeat at Fredericksburg, Burnside noted that had the pontoon bridges he requested not been delayed, the Union forces would have won this battle, and the lives of those soldiers lost on the battlefield would have been spared. Burnside hoped to repair his reputation with a second attempt on Fredericksburg in January. As he tried to move his troops into position, a torrential rain created a quagmire of the roads to the new site of the crossing. Burnside ended up abandoning the "Mud March." The next day, January 26, 1883, Burnside was replaced with Major General Joseph Hooker. Two major generals of the Army of the Potomac—McClellan and Burnside—had now been relieved of their duty due to the catastrophic delays that characterized their terms as commanders.

Three months later, Joseph Coryell would be dead, shot on the shoreline of the Rappahannock. After I visited the railroad bridge in Fredericksburg, I drove down Route 3 to visit the site where he was shot. I made the journey on a warm February day, with full sun and warm temperatures melting the snow that had fallen the week before. Just beyond the farm where George Washington spent his childhood, there was a small road off the highway that led down to the site where the 24th Michigan crossed on April 29, 1863, the crossing that marked the start of the Chancellorsville Campaign. I nearly passed the small turnoff, but saw it at the last second. I pulled onto the road and was immediately confronted with a large red gate and several "No Trespassing" signs. The land where the army camped out, once on Henry Fitzhugh's estate, was now property where, the signs warned, "trespassers will be prosecuted."

On the south end of the property stands a water-treatment plant, along with a public boat-launching site. I walked down the road to the boat launch, whose wood pylons I used as steps onto the shore of the river. I hoped that I could simply walk along the shoreline until I reached the site of Fitzhugh's Crossing, where the Iron Brigade attempted to build pontoon bridges into Fredericksburg to initiate the Chancellorsville Campaign. As I stepped onto the shore, my shoe sank about six inches into the mud. The tide was out and the shoreline was thick, impassable mud.

On my way back up the road, walking with one mud-covered shoe, I ran into a man in his late thirties unloading groceries from a minivan in front of his house. I was desperate to find a way onto the site and asked him whether he knew anything about this property and its owner. "Yeah, that's my land, actually," he said, holding a jug of milk and some oranges. The back hatch to the minivan was open, groceries sitting in brown bags, his two-year-old son asleep in a car seat. I told him I hoped to tour the property to see the site of Fitzhugh's Crossing and the place where Coryell was shot. "What's the point of touring the land?" he asked. "I don't think you're going to find anything there from the Civil War. That land was used as a mining facility starting in the 1970s, and the whole property was dug up. It's really rough land." I told him that part of my work to understand what happened to Coryell was to walk these sites. Place is an important part of knowledge; place tells stories in ways that are otherwise inaccessible. "I'm not sure that I feel comfortable with you walking that property by yourself. The land hasn't really been touched since the mining company left, and there are hazards everywhere."

This man's family had farmed this part of Virginia since well before the Civil War. The house we were standing in front of, which he recently had renovated, goes back several generations in his family, all of whom were farmers. His Virginian ancestors worked on the farm owned by Mary Washington, just across the street from us. Behind his house stretched acres of farmland, now converted into a conservation site and a place where chickens freely roamed, laying eggs that he sold. This location meant that his family had been neighbors to the Fitzhugh property during the Civil War, and now he owned the land that was once the site of a crossing that began the Chancellorsville Campaign. "I tried to grow crops on that land, but they won't grow." Now he wasn't sure what to do with the land. "You're only the second person to ever ask me about the land. No one talks about that part of the battle. The re-enactors never come this way when they're re-creating the Civil War battles."

We looked at the property on Google Maps, with the satellite view turned on. There was a road that led through most of the property, and Fitzhugh's Crossing had been right at the bend in the road. "That part of the property is safe," he said. "The large trucks used it when they were hauling the stone and sand off the site. If you stick to that road, you won't run in to any hazards. There's a clearing right there that will lead you down to the river." With that, he gave me the go-ahead to walk the property.

I pulled onto the same small road that I had visited earlier that day and parked my truck off to the side. I put on an orange surveyor's vest (not wanting any trespassing hunters to mistake me for something worth shooting) and began walking down the rocky, unpaved road. The size of the property surprised me; it was vast and incredible and peaceful. Deer scattered quickly once they saw me, undoubtedly surprised to see a human being on this land. I kept to the road, keeping an eye out for any deep holes or stray equipment that the mining company might have left behind.

I came to the clearing next to the river, the site where the 24th Michigan camped out in preparation for their crossing on April 29, 1863. Just before arriving at this spot before dawn, Joseph Coryell wrote his last letter home. In it, he spoke of getting paid and sending only part of the money back home "for if it was lost, it would be all lost." He tells Sarah of a recent robbery of soldiers' letters home that netted the thieves more than forty thousand dollars. He ends by saying, "You must see that the mortgage is properly discharged when you make the last payment. The order has just come around to march at twelve o'clock today so I must be packing up. Goodbye my Dear wife. May God bless you. From your ever affectionate husband, Joseph Coryell."

As the Iron Brigade attempted to cross the Rappahannock, the scene began to look like the first day of the Battle of Fredericksburg all over again. Under dense fog, in the dark of the early-morning hours, bridge builders scattered under fire from the opposite side of the river. One soldier in the 24th Michigan wrote, "The enemy's fire become so hot that the engineers and train guard had to leave the boats and fall back."[14] The sides exchanged fire for some time, and then the Iron Brigade was ordered to fall back to a safer site just on the other side of the ridge that led down to the river. As the men fell back, a bullet struck Joseph Coryell in the head. He fell on the shores of the Rappahannock next to the supplies for the pontoon bridge. He was put on a stretcher and was carried just over a mile up to the Fitzhughs' house, called Sherwood Forest, which had been taken over by the Union Army as a field hospital. He died the next day and was buried in the nearby orchard, where the 24th Michigan held a funeral for him.

I sat on the banks of the river, looking across where the soldiers eventually established their bridgehead. The location was now occupied by large houses with docks, sundecks, and platforms with rope swings into the river. Just in front of me was a decaying metal and wood barrier used in the 1970s to keep boaters from docking on the mining company's land. One of the huge houses across from me had a tall flagpole flying an American flag; nearby, a homeowner had tacked a Confederate flag in a window.

I made my way back to my truck and passed hewn stones carved out of the land by the mining company. They reminded me of headstones, and I was glad that Coryell had not been buried here only to be accidentally dug up by miners who might have no knowledge of the battle that took place on this property. Up the road, on Fitzhugh's estate, the house was boarded up and rotting and the land had been purchased recently by a developer who will probably build a new neighborhood of homes.

I pulled back on to the highway. It felt strange to leave this location with the thought that Coryell never did. The chaplain of the 24th Michigan, in his letter to Coryell's wife telling her that her husband had been killed, sent along several letters that Joseph hadn't had a chance to put in the hands of the postmaster, including his final letter reminding her to pay the mortgage. Already in transit was her final letter to him, arriving days too late. In it, she apologizes for not writing more often. Their son was sick with teething, and tending to him—and to the farm—had been exhausting. She talks about the body of one of the local soldiers returning to DeWitt, and describes the large funeral the community held for him. She can't bear the thought of more young men dying, which seems inevitable to her. Her letter was never opened once it arrived with the Army of the Potomac; instead, it was sent back to Michigan, arriving back to the DeWitt Post Office shortly after Sarah received word that Joseph had been killed.

I pass by the railroad bridge again on my way out of Fredericksburg. A bright orange and white pole juts out of the ground, warning people not to dig here, where fiber optic cables have been laid. While waiting at a red light, I text my wife to tell her that I am on my way home. I see the cell tower near the road that my phone is probably connected to at this moment, my signal hitting the antenna and then running down the lines and out to Verizon's mobile switching center. On my screen, my text message to her is marked "Delivered," then "Read." She doesn't respond right away.

The text message I sent allows her respond when she gets a chance. So while messages are moving at the speed of light along these pathways in our own era—compared with letters traveling these routes during the Civil War, which typically took over a week to journey to Fredericksburg from DeWitt, Michigan—we still end up spending much of our social lives waiting in ways that resonate with the soldiers of the nineteenth century. Waiting was a part of the messages sent back then, and the same can be said about the messages sent along the fiber optic cables attached to the side of the railroad bridge in Fredericksburg.

The notification that my wife has read my text message is feedback to me, verifying the delivery of my message. Once I get home, I'll pack another suitcase for a flight to London, where I'll be spending time in the National Archives. While there, I'll see some of the first "read receipts": medieval wax seals. These seals were also used as marks to authenticate the author, similar to the signature at the bottom of every one of Joseph Coryell's letters home. These seals were a mark that gave people a way to confirm their identities across the gap to distant contacts. The marks, which traveled across networks of uncertainty, reveal the power and the problem of distance.

6 MARKS OF UNCERTAINTY

I was standing in a long row of archival boxes in the British National Archives on the outskirts of London. I had been escorted through several sets of alarmed doors to the stacks where the archival documents are held. I was then led to this row of boxes; most had the letters HCA stamped or written on the side, which stood for the High Court of Admiralty. Amanda Bevan, head of legal records at the National Archives and the lead archivist working on these so-called Prize Papers, opened one of the boxes and pulled out a charcoal-colored mailbag. Inside were stacks of unopened letters. "Back before Post Offices, if you wanted to send a letter to a distant relative or business contact," she told me, "you'd hand it to the nearest ship's captain that was heading in the right direction." Often in pubs or coffeehouses, these mailbags would hang near the bar; a local would drop his letter into the bag of the ship heading toward the recipient. Yet none of the letters in this collection was ever delivered. Most were still sealed shut. "This mailbag, for example, was confiscated by the Royal Navy during wartime from a Dutch merchant ship. When a ship like this was captured, the mailbag onboard was one of the items plundered. The mail was used as evidence in the High Court of Admiralty." She carried the archival box with the mailbag over to the worktable where her team has been painstakingly sorting through this corpus of letters for five years. A colleague opened a sealed envelope and pulled out some handmade glass-bead bracelets and a letter in Dutch. I asked Bevan how many letters were a part of the Prize Papers. "I think it would be an underestimate to say 150,000 letters," she responded. "What you have here," one archivist told me, "is the equivalent of an early-modern Post Office being invaded." In every one of the archival boxes in that long row were hundreds of letters that had never arrived at their destinations.

Many of the letters were closed with a wax seal that created a mark unique to the letter writer. I had come to the British National Archives to study sealing practices of medieval letters, especially those used by kings, queens, and others in high positions in society. Wax was often used as an adhesive to seal a letter that was meant to be in confidence (called a letter close). The seal simultaneously functioned as a marker of the person sending the letter and as proof that the letter had not been tampered with. Many of the letters in the Prize Papers archive—written centuries

Figure 19. The middle of the long row of archival boxes containing the High Court of Admiralty (HCA) Prize Papers in the British National Archives. These letters represent the hijacked correspondence on board enemy ships captured by the Royal Navy. Image © 2017 Jason Farman.

after the first English sovereign used a seal on letters—remained sealed and undisturbed. The seal did not deter a letter's interception by plundering naval troops, but it was now a mark of failed communication.

Seals initially served the practical purpose of functioning as an adhesive to shut the letter securely and mark the identity of the author on the exterior of the letter. In addition, the seal served as verification to the recipient of secure delivery. These documents were typically rolled shut, with a string wrapped around the parchment, and the mixture of beeswax and resin impressed over the strings. The recipient opened the document by first breaking the seal. Because damage to the seal was part of its function, few broken seals survive; most were destroyed or discarded after the letter was opened.

This practice of sealing a letter soon changed: King Edward the Confessor decided that his seal should remain intact after the letter had been delivered as proof of his authorization of the document. Instead of using the seal itself as an adhesive to close the letter, Edward began attaching his seal to a strip of parchment that hung from the bottom of the document. A knife was used to cut slits in the parchment, and the strip was woven between these cuts so that it came together where the wax was poured. The seal was then impressed over this strip, serving as a visual link between the document and the person writing it. Alternatively, the very bottom of a document might be partially cut or torn into a long strip that hung from the document, creating a place for a pendant seal to be hung from the parchment.

In both early and later form, seals represented the power and the problem of distance. There is tremendous power in our ability to send messages across vast geographic distances. Messages allow us to keep in touch with distant contacts. They afford the opportunity for business contacts to maintain a relationship and carry out their dealings from remote locations. Long-distance communication allows people to share ideas and knowledge from different cultures around the world.

This kind of communication is problematic, however, in that people are not communicating face-to-face. People's bodies are

Figure 20. The great seal of King Henry I attached to a document by a strip cut away from the bottom of the parchment. This document, written in A.D. 1115, confirmed the transfer of land to the Church of St. Mary in Wix, Essex, England. Image © 2017 Jason Farman. Used by permission of the British National Archives, ref. E 42/316.

removed from the act of communication and are instead attached symbolically to the document that is being sent. The seal represents one approach to this problem: It serves as a symbol of the hand that stamped it, both metaphorically and literally. A seal often incorporated imagery of the person who owned it, and the act of making a seal impression linked the human body and its intentions to the document that body sealed.

A saying is attributed to the English king Alfred the Great in the ninth century: "Consider now, if your lord's letter and his seal came to you, whether you could say that you could not recognize

Figure 21. The privy seal of the Dutch countess Elizabeth Rhuddlan, from around 1300. This seal shows the countess standing, flanked by two coats of arms. Many seals displayed images of their owners and were simultaneously understood to be impressed by the hand that made the document. Image © 2017 Jason Farman. Used by permission of the British National Archives, ref. SC 13/F150.

him by this means."[1] The seal was a symbol of the letter writer that would be instantly recognizable by anyone, even the illiterate. The link between seal and imprinter was readily visible. Kings and queens typically had double-sided great seals, with one side often showing the sovereign on a horse as a knight or warrior, displaying military might. The other side showed the ruler on the throne, often holding a staff and orb. The massive size of these great seals also communicated the importance of the sovereign. As I sat at my reader's desk in the British National Archives, I was presented with a document that had Queen Elizabeth I's great seal attached to it. The seal, as I lifted it from the box that protected it, was as large as my fully extended hand. These great seals required a significant amount of wax and were operated by a mechanism similar to a printing press, with metal matrices simultaneously pressing the designs on both sides into the wax.

The uncertainties of geography, including the time lag that distance enforced, created a culture that demanded particular symbols as verification of the body on the far side of these gaps. The seals and undelivered letters in the Prize Papers highlight the uncertainty of long-distance communication once documents began to be sent around the world in abundance. The waiting that was part of the process of sending documents represents the precarious nature of communicating in an era before official post office departments and international infrastructures for message delivery. Without dedicated routes and infrastructures in place, delivery of letters required creative combinations of commerce, transportation, and ad hoc messengers. The resulting long-distance social networks were founded on uncertainty and uneven expectations of reply. When we send messages today, reciprocity is a normal expectation. As I discussed in the introduction, reciprocity is a key to feeling connected with one another, as each response to a message confirms a back-and-forth that links us with the people in our lives. However, when the infrastructures and pathways (ships, mailbags, trade routes) made a reply uncertain, how did that shape communication? How does the uncertainty of response change the ways that letters were composed? How would the experience of waiting change if the expectation was that a

Figure 22. Holding the great seal of Queen Elizabeth I. The great seals of English sovereigns were massive and double-sided. This side shows the queen on a horse, symbolizing her military power. This was a trope that had been displayed for every English queen and king who had a great seal. This seal is attached to a strip of parchment at the bottom of a document dated May 22, 1564. Image © 2017 Jason Farman. Used by permission of the British National Archives, ref. E 30/1146.

response might never come? What happens to waiting when a delay could be infinite?

The rise of documents and the spread of messages between people not of the nobility or clergy required many social changes, including the rise of a literate public, increased availability of the tools for writing (ink and parchment or paper), and a method of exchanging these messages. The rise of documents simultaneously required a massive cultural shift from honoring oral traditions to honoring the written word. "Documents did not immediately inspire trust," says Michael Clanchy in his seminal book *From Memory to Written Record.*[2] Though writing existed in Western culture for thousands of years, it had consistently been understood to be a lesser form of communication than the spoken word. In British medieval society, oral testimony was well established, dependent on a Socratic tradition in which the spoken word was a direct line into the mind of the person. In contrast, a letter could not be cross-examined or engage in dialogue. Media scholar Walter Ong writes, elaborating on the work of Clanchy, "Witnesses were *prima facie* more credible than texts because they could be challenged and made to defend their statements, whereas texts could not (this, it will be recalled, was exactly one of Plato's objections to writing)."[3] Official documents also initially lacked such standards as the inclusion of a date, signatures or seals, or notarization by witnesses. They couldn't easily be authenticated, and forgeries were not uncommon. It was more efficient to disprove someone's oral testimony about a conflict over land ownership, for example, than it was to trace the provenance of a charter.

Yet the power of documents eventually overwhelmed these prejudices. The document allows the letter writer to be effectively present in different places at once and to exist across time, beyond a single lifetime. As connections became global in scope, connecting cultures and societies around the world, documents served as the nervous system of that network. Scholars like Clanchy and Ong rightfully point to the intimate link between commerce and documents; the need to create a record of transactions and ownership was a major driver behind the establishment of documents. The spread of messages around the world was logically and practically related to trade routes across ocean or land.

As we saw in chapter 3, a lag in internet connection directly relates to lost revenue for companies like Amazon. Commerce and time were no less connected in the past. The duration of waiting for response to message, linked primarily to the speed of transportation, affected the exchange of goods. This principle was described to me by Matt Greenhall, a scholar of sixteenth- and seventeenth-century economies and trade, and an archivist at the British National Archives. His work details the impact of the plague on trade practices during this era. Not only was the timeliness of messages linked to the ability to conduct trade, but delays could also create life-or-death scenarios for those onboard some of these trade vessels. Greenhall points to the great plague of 1665.[4] The plague originated in Holland, and any ship arriving in England from a Dutch port was quarantined. Often the items aboard would be perishable, but port officers nevertheless required verification that the departing vessel had left before the plague had hit Holland, or that it had not been in contact with anyone from a port city known to have the plague. Sailors aboard the vessel also had to sit and wait until confirmation arrived about their journey dates and their freedom from contact with plague-infested ports. This meant that these crews would run through supplies, including food and fresh water. Beyond the health of the sailors, the delays put their goods at risk and delayed the conveyance of the mail onboard the ship. Again, these ships were the de facto postal service at the time; so while the sailors aboard waited for a letter of clearance to engage in trade, those who were expecting a letter borne by this ship also had to wait.

Plagues, war, and seizure of mail from ships created uncertainty for these networks of communication. One principle has resonated across eras as networks have been established to send and receive messages: The political and cultural contexts that surround communication networks directly influence the speed at which messages travel. Networks are never neutral pathways that allow messages and data to flow across their channels; instead, the speed or delays that affect message exchange are often direct extensions of how these networks are influenced by the sociopolitical landscape of the time. From the uncertainty created in the early modern era by war and other disruptive events to our own uncertainties developed out of attempts to undermine net neutrality, the flows of messages across the dominant network of an era are tied to the ways that the network takes shape around conflicts, crosscultural divides, and global events like plague or earthquakes.

Owners of seals knew very well that message networks were not neutral and that these networks do not treat all messages equally.

Instead, though messages limited the seal owner's ability to exert influence face-to-face, seals were a way to exert that influence through markers of their status. Seals allowed the sender to create a visible marker of status in order to project influence along with the message across the divide to its recipient.

In the first centuries of sealing in England, the owner of a seal had some kind of status or authority. The sealer tended to be literate and to deal in business of one kind or another. King Henry II's chief justiciar, Richard de Lucy, said, "It was not formerly the custom for every petty knight to have a seal, which benefits only kings and important people."[5] Clanchy has noted that having one's own matrix for creating a seal implied status and the seal owner's ability to interact and understand the complexities of documents, "even if this were an aspiration rather than a reality."[6] These seals were marks on a message that impressed upon the recipient the prominence of the one sending the message and the implication that a response should not waste the valuable time of this important person. Power, privilege, and expectations of quick responses were bound together in these seals, affecting how people in different social standings experienced waiting.

This use of seals to exert influence was especially important for women of the medieval era, who maintained little power but could often exert considerable sway over the society. Elizabeth Danbury, a scholar of seals and medieval documents, invited me to the Society of Antiquaries of London to look at their collection of seals, many of which are women's seals from the Middle Ages. Danbury slid out a long, thin drawer that displayed seals that had been collected by the society over the past century and a half and took it to a nearby table. "Many women would use the iconography on their seal as a reference to the men in their lives. They would sometimes use heraldry of their husbands or sons to position themselves in relationship to the power of the men in their lives." In contrast, though, she pointed out the great seal of the twelfth-century English claimant Empress Matilda, a seal that was much smaller than the one Queen Elizabeth I would use four hundred years later, and slightly smaller than her cousin Stephen's great seal. Matilda and Stephen fought over the crown, and Matilda was never recognized as queen of England. Eventually her son, Henry II, claimed the crown. Danbury described the inscription on Empress Matilda's grave, which says, "Great by birth, greater by marriage, greatest in her offspring, here lies daughter, wife, and mother of Henry."[7] Empress Matilda, though she could easily draw on the men in her life as a rightful lineage to some semblance of power, instead is seen in her seal, seated in the throne, holding an orb in her left hand and a scepter in her right. Here Empress Matilda was following the expectations of the great seal of a sovereign, and, as Danbury notes, she found "a way to express [her] personal achievements, aspiration and influence" through her seal.[8] Her seal conveys her power and thus sends a very different message from the one inscribed on her grave or the seals of other women who identified themselves through the men in their lives.

I had seen Empress Matilda's seal earlier in the day at the British National Archives. In order to see the seal, I had to request it through the archive's online catalogue, which bears a note specifying that the document must be viewed in the "invigilation room," a term I later learned refers to close observation. I requested the seal and document and made my way over to the locked invigilation room, where my bags were examined and I was buzzed in. The box was brought in for me, and I sat at a chair with several surveillance cameras overhead. (This experience was in vast contrast to my interaction moments earlier with King Henry I's great seal from A.D. 1115, which I had examined freely at my desk in the reading room of the archive. The seals restricted to the invigilation room are the archive's best specimens, intact seal impressions, some of them the only remaining version of a seal.) I picked up Empress Matilda's great seal, from A.D. 1141, and turned it over. While the front of the seal was pale in color, the back was a brownish red.

Unlike other great seals, which were imprinted on both sides by a matrix, Matilda's seal was imprinted only on the front. In previous books and digitized images, no photograph of the back of Empress Matilda's seal is included. This is true of almost all the seals that I studied: The obverse (or front) of the seal is photographed or digitized, but the reverse is not unless it contains an impression. What interested me about the reverse of many of these

Figure 23. The great seal of Empress Matilda, once attached to a document from A.D. 1141. The reverse side of the seal shows several fingerprints of the person making the seal impression. Images © 2017 by Jason Farman. Used by permission of the British National Archives, ref. DL 10/17.

seals is that they often contained marks of the body that sent them. Here, for example, were a number of fingerprints across the back that had been impressed into the wax when the seal was made.

Marks of the body are common on seals, though only sometimes were they intentionally included as a marker of the person who sent the sealed document. A medieval rhyme about William the Conqueror claims that he sealed his documents using his teeth (his "fang Tothe," to be precise) instead of a seal matrix. In another case, in the mid-fifteenth century the count of Lincoln sealed a grant with his teeth marks instead of the impression made with a matrix. Other seals I looked at contained knuckle indentations or deep fingerprint impressions in the back. These marks, whether intentional or incidental, capture the essence of what a seal was meant to symbolize: the presence at the document of the body of the person sending a message. Brigitte Miriam Bedos-Rezak, scholar of seals (a field known as sigillography), has noted that seals "embodied his person as the true originator of the act in question—his presence often rendered even more tangible by the inclusion of bodily marks in the seal, such as fingerprints, bite marks, or actual hairs plucked from his beard."[9]

This presence is key because the uncertainty inherent in sending messages interferes with the expectations of face-to-face communication. Messages eliminate many nonverbal cues, the debate and back-and-forth that allow people to expand on knowledge, and the trust established by eye contact and bodies in close proximity. Moreover, the body of the person represented the site of power. A signature is evidence of a person's presence at a particular place at a specific time. The presence of one's body signals the ability to authorize a document. So, too, did the mark of the human body and a person's authorization given through a seal impression.

The seal, however, was a mechanical technology that allowed the same impression to be reproduced over and over. This meant that anyone could use a seal matrix on someone's behalf, and the mark itself would look identical (though the fingerprints on the back would be different). So while the seal matrix was often carried by the owner, sometimes even hung around the neck like a necklace, there were times that the seal was removed from the body that it was meant to represent. For example, when a king traveled, a "seal of absence" could authorize actions on the absent king's behalf. Later in the history of sealing practices, the invention of signet ring seals allowed the owner constant bodily possession. These seals were in contact with the body of the imprinter more intimately than any other, crafted for the fingers of the sealer, worn at all times.

As media theorist Walter Benjamin has noted, the act of mechanical reproduction creates a troubled relationship with an object's "aura," or the status that tends to give an object its power and rarity.[10] A painting, he notes, has an aura because the artist touched this specific canvas. The paintbrush he or she held made contact with the canvas, which is the only one of its kind. The viewer can experience the presence of the artist, sometimes even by seeing his or her brushstrokes across the painting. The aura of the painting resides in these rare features that cannot be reproduced. Benjamin then asks where the aura resides for something like a photograph. Which is the original? Is it the first image produced by the photographer? Is it the negative? A painting's value, in some regards, is in its uniqueness. Reproduction, like the work of Andy Warhol, for example, disrupts the traditional connection between the artist, the artwork, and the idea of the original. This kind of troubled relationship between a person's body and the thing produced, undoubtedly, was in the mind of those who encountered seals. Beyond forgery—which was a problem with seals that could be reproduced—this ability to replicate something that was detached from the author's hand created a quandary over the aura of the seal.

In addition, the political or economic utility of most seals was a Christian context to the cultural understanding of seals. An impression was made in the image of the maker, mirroring the Christian belief that mankind was made in the image of God. The image of the sealer was often on the seal design itself, and the seal was an impression made by the author. Just as the believer considers herself the imprint of God, the seal is an imprint of its author. Thus the practice of sealing could be understood to be a Godlike act of creation, imprinting identity onto the document.

The seal gave the sealer the sense that his identity was going out into the world on his behalf, through the impressions he had made on the document.[11]

As Bedos-Rezak mentioned to me, there has always been a relationship between identity and reproducibility: The documents we sign are seen as "authentic" because they match the signatures we have produced time and time again. Linking identity to a document (like a love letter or even a check you sign and send through the mail) depends on the ability for others to recognize the signature based on its close match to a repeated form. From Bedos-Rezak's perspective, reproducibility is the "condition of authenticity." Reproducibility, she argues, isn't the opposite of the real and authentic; instead, the aura actually depends on our ability to trace features in a document that can be replicated and verified.

Seals therefore not only represented the body of the sealer but also linked the abstract document to the reality it was signifying. The authenticity that was associated with seals was found in this link between the person's body, the act of making an impression, and the Christian notion of something being made in the image of its maker (of breathing life into the dust, as it were). This progression is how the term "character" came to mean upstanding moral identity. "Character" initially meant a seal that labeled a thing with its intended purpose. Later the term became a theological one, "to designate the foundational imprint made on the Christian by baptism, and by those other sacraments—confirmation and ordination—which, once applied to an individual, marked him indelibly for the service of God. Character, in this sense, once impressed, was an absolute certainty."[12] Notions of impression, of making something in your image, gave authenticity to the reproducible act of sealing a document.

Within this early Christian context, the certainty of impression—as an act that linked the identity of the letter writer to the document through the seal—was one way to offset the uncertainty of long-distance communication. The outcomes of sealing practices, however, reached beyond these intentions to shape the very nature of the society. The uncertain networks across these distances created, in part, a culture that used particular symbols as verification of a person's identity across these gaps. These symbols, in turn, created a literate culture.

Seals played an important role in the move from oral to print culture. As documents became a common part of daily life, seals began to be used by anyone who had business dealings, not just those with wealth and power. As a result, there was a bridge built between literate and oral cultures. Gradually oral culture evolved to incorporate the symbolic representations of seals, en route to full literacy. Clanchy makes a persuasive argument along these lines when he notes that documents were a requirement for widespread literacy. As with technologies today, early adopters were those with means, yet the expansion of the technology of documents and seals meant that they began to make an impact on the common life of those in a shire. "The gentry were not going to learn to read until documents were available and necessary," Clanchy writes.[13] This extended literacy and availability of the tools of document creation have a sweeping impact. This mode of literacy eventually led to what Marshall McLuhan has termed the Gutenberg Galaxy of print culture. This shift, for McLuhan, creates a massive transformation as the sensory engagement with communication: first, the ear listening to the spoken word, then the eye looking at the page of the printed document.[14] Literacy, as signaled by seals and documents, transformed the mental map of humankind and the ways that we encounter the world.

The transformation wrought by the printing press required a literate culture familiar with documents. This all emerged, in part, out of the bridge technology that seals provided. Seals, as the bridge between oral and literate cultures, fundamentally changed how people interacted with information and documents. These tools also changed how people interacted with one another, as they were able to send textual messages to distant contacts. This ability expands both time and space, as information is stored in new ways and people are able to connect with contacts far beyond the reaches of their small shire.

As letters and other documents became part of everyday life, the routes to deliver these documents needed to become less precarious.

Sometimes people waited for a response because the available delivery methods were slow, but sometimes their messages—or the responses to them—were intercepted. As represented by the roughly 150,000 letters in the HCA Prize Papers, having a letter intercepted was not an uncommon event.

While seals functioned as an authorizing mark on a document that the hand of the person writing it was indeed the stated author, they also initially served as the mark that a "letter close" had not been intercepted and read (or changed). A broken seal was a visual cue that the contents of a document should no longer be considered private. A visual mark like this carried over into other forms of intercepted correspondence throughout the history of message exchange. These marks were important since the likelihood of interception increased as transit times increased. As people waited, they become more uncertain about the delivery of their letters.

Concerns about privacy and secrecy are characteristic of these uncertain networks of exchange. Once the message departs the body of the sender, the sender has no ongoing control over the process of delivery. This middle ground between sending and receiving a message is where "surveillance from the middle" happens; that is, interceptors are mediators able to engage in surveillance over these documents as they are in transit.

Sometimes, the evidence of interception, though circumstantial, is that the wait for delivery never ends. Letters never arrive and replies are not sent. Other times, as in the case of a broken seal, evidence of interception is visual. I saw a later example of visual evidence during archival work with wartime correspondence. During World War I, Walter Boadway was a cadet aviator stationed in France and other parts of Europe. He wrote often to his wife, Betty, who lived in Pasadena, California. As Betty began receiving his letters shortly after he was deployed, there were strange holes in the paper; sometimes entire portions of a page seem to be missing from his letters. After this happened several times, she wrote to him to let him know that words and paragraphs from his letters were being removed for some reason. These parts of letters were not being blacked out by military censors; instead,

Figure 24. A letter from Walter Boadway written during World War I, with a word removed by a military censor before his wife received the letter in Pasadena, California. This letter is a part of the corpus of letters owned by the Center for American War Letters at Chapman University. Image © 2015 Jason Farman, reproduced courtesy of the Center for American War Letters.

the censors were using scissors or razors to literally remove any word or phrase that might compromise classified information.

For Boadway, what constituted classified information remained mostly unknown, since the definition was constantly changing in response to wartime circumstances. In his early letters to his wife, he had no sense that naming the kind of plane he flew was exposing unknown information that, if intercepted, would compromise the military's mission. He was also told at one point that he could not date his letters; instead of writing "November 11, 1917," he simply wrote "Sunday." Otherwise, the date would be cut out entirely by the military censors. As he wrote these early letters, he used both sides of a sheet. After the military censors finished cutting out classified information, his letters were sometimes mostly illegible. The possibility of an intercepted (and censored) letter required that letter writers transform the ways they wrote.

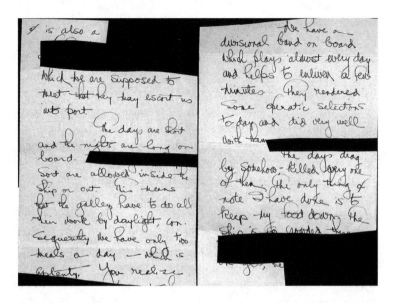

Figure 25. Two sides of the same censored sheet in one of Boadway's letters to his wife. This letter is a part of the corpus of letters owned by the Center for American War Letters at Chapman University. Image © 2015 Jason Farman, reproduced courtesy of the Center for American War Letters.

As these words, sentences, or paragraphs are erased through the removal of the material piece of paper, the system of interception is revealed. Here, much in the way the broken wax seal on a letter revealed that the letter had been read, the censor's cutting of the letter was evidence of interception. For soldiers, writing words that were prohibited resulted in a diminished ability to communicate. But the removal of the words revealed the flow of communication: from author to censor to intended recipient (the censor guarding against a flow sidetracked to enemy intelligence). In the process of surveillance in the middle, the act of interception is often evident at the endpoint of communication.[15]

When people become aware of the consequences of interception, they engage in tactics to avoid those consequences. It took Boadway some time to adapt this correspondence to censorship. Initially, he simply tried to adjust the content of the messages

without considering that removal of a word on one side of the
paper would also remove part of the (probably innocuous) message
written on the other side. He soon started writing on only one side
of every sheet of paper, absorbing the increased cost of postage
from France to California. As the military learned of ways that
enemy intelligence was intercepting communiques of strategy,
movements, and troop strength, it censored its own soldiers ac-
cordingly. When Boadway learned of ways that his letters were
being censored, he responded in turn, writing on only one side and
avoiding language that would concern the military censors. His
response to censorship (as one mode and outcome of surveillance
in the middle) was to tactically respond to the surveillance strategies of those monitoring his letters.[16] Similar to the ways that
Boadway adapted his letter writing for the uncertain gap between
sender and receiver, seals emerged as a method of dealing with
this intervening space and the wait times that accompany these
precarious networks.

Reciprocity, or the mutual exchange of items like letters, was his-
torically among the few ways of confirming the receipt of a letter.
The unbroken seal was a sign that the letter had arrived undisturbed
and was a receipt of successful transit. To receive a letter back
typically confirmed for the sender that the initial letter had arrived.
Unbroken seals on "letters close" in both directions were receipts
of successful mutual delivery. For wartime letters, as we saw in the
previous chapter, writers would often begin by creating a correspondence timestamp such as, "I received your letter of September
22 yesterday." Until a letter was returned, the wait time was a
period of uncertainty about the status of the communication network. Would the letter or document arrive? Would it arrive without
first being intercepted?

Yet one can withhold reciprocity as a power maneuver; by
keeping a correspondent waiting, one can hold the other in a state
of uncertainty. One famous example occurred on July 13, 1530,
when a letter arrived to Pope Clement VII asking him to grant
dispensation to annul the marriage of Henry VIII to Catherine of
Aragon. Affixed to the huge document were eighty-one seals, one

from each of the lords and religious officials who signed the document, threatening the pope that if he was declined to dissolve Henry VIII's marriage, they would be "left to find a remedy elsewhere." Henry VIII, seeking to marry Anne Boleyn and produce a male heir, had ordered the letter written and sent. The seals affixed to the letter served as visual symbols of the importance of the lords and archbishops and their ability to circumvent the pope should such actions prove necessary. Henry waited for a response, but the pope's delayed response took too long, and Henry demanded an act of Parliament to sever ties with the Roman Catholic Church. He assumed the position of "Supreme Head of the Church of England." The pope was using delay as a tactic to emphasize his power; however, this strategy only infuriated Henry VIII, who decided he was too powerful to wait any longer.

Tactics of delay cause uncertainty and dismay in our own era, as well, as when message receipt is displayed on the text messages exchanged on mobile devices. Though the network no longer carries the same uncertainty of early message networks, the waiting associated with mobile technologies creates similar feelings of uncertainty that shape reactions to messages. Text messages marked as "read" are verified to the sender. By opening a text message, the recipient in essence "breaks the seal." Once that occurs, some platforms provide immediate feedback to the sender. Once the message is read and its receipt verified through a read receipt, a longer-than-expected wait time for response can create uncertainty and misunderstanding between correspondents.

In Japan, two terms highlight these feelings of uncertainty as they relate to waiting and reciprocity: *kidoku-through* and *midoku-through*. These terms are applied to the read receipts that are displayed on such messaging platforms in Japan as the popular LINE app that most young people use. *Ki* means "already" and *doku* means "read"; so *kidoku-through* (borrowing an English word) means that the message has already been read through. *Mi*, on the other hand, means "not yet." So a *midoku-through* message has not yet been opened. The seal remains intact, to draw the parallel to the medieval period. Mobile users in Japan who use the app LINE to send text messages—about 95 percent of all smartphone users in the country—do not have an option to turn off read receipts. So every single LINE user knows whether his or her message is *kidoku-through* or *midoku-through.*

I met with executives and designers at LINE in Tokyo to discuss their use of read receipts. LINE was created in the immediate aftermath of the Tōhoku earthquake as a response to the uncertainty of message receipt that I discussed in chapter 1. One of the key aims for the app's creators was to include a feature that let the message sender know when the recipient opened a message. In practice, though, once the seal has been broken on a LINE message—once it becomes *kidoku-through*—the waiting expectation shifts and a prompt reply is expected. To leave your correspondent waiting is either a sign of disrespect or a maneuver to communicate power to the sender. Messages that are not immediately responded to can communicate a perceived differential value of time. That is, a lack of a quick response communicates, "My time is more valuable than yours and I will respond when I get a chance."

Waiting is often a result of geographic distance; people are far apart from one another and the only mode of communication they have requires significant wait times. These wait times, due to this distance, create a back-and-forth that seems to efface the bodies of the people communicating. Seals are designed to reinsert the body into those exchanges, marking and authenticating the documents being exchanged when people are unable to be together. Since the time of sealing practices, we have continued to require authentication in an era where much of our interactions are done remotely. We live in a world that privileges the message, and messages continue to require proof of authenticity. These range from passwords to two-step authentication that allows us to access our email. We still sign important documents and link our communication with our bodies in several ways. We memorize passcodes and unlock our mobile devices with facial recognition or fingerprints.

Seals have not disappeared altogether; in fact, they are used widely in countries around the world. In the United States, we have seals that authenticate important documents like birth certificates. Each state in the United States has its own seal, and these are used on legal documents and official pieces of legislation. Seals still carry quite a bit of importance for such organizations as universities and cultural institutions. Victor Mair, a professor of Chinese language and literature at the University of Pennsylvania, recounts a story of the cultural importance of seals in China. He was working with the director of an archeological institute in the country, who was offering to transfer some artifacts to Mair. In order for Mair to carbon-date the artifacts, he needed transfer of the pieces to be authorized by the director, confirming that he had indeed assigned ownership. "He said he'd be happy to provide such a certificate, but that his institute had undergone a reorganization and renaming, and he was waiting for a new seal," Mair said. "So I waited patiently for the certificate, but the seal just kept getting delayed. Finally, when it came to the point that I desperately needed the certificate right away, I told the Director just to send it to me with his signature affixed." In response to this request, the director laughed: "Such a certificate would be worthless!" To the director, a signature meant nothing and the seal meant everything. The director finally agreed to simply sign the certificate and later was shocked that the signature was accepted and was more important than the seal of his institution.[17]

Seals are common today in many Asian countries and are used by most citizens to sign their names on official government documents. Every person I spoke with in Japan has a seal. Called a *hanko* (or a more formal version, the *inkan*), this small seal matrix is required of all Japanese instead of a signature. The seal is carried by its owner at all times and can be used to authorize documents, used with ink rather than wax. This practice started in the 1870s, as a growing bureaucracy in the country required that documents be authorized. In what should now be a familiar trajectory, before this time, only people in positions of power had access to personal seals. Now, anyone can buy one at a neighborhood shop, and they range from a modest device with a plastic handle to an extravagant version incorporating materials like ivory or precious metals. The average person in Japan has four to five personal *hanko* in his or her lifetime, and some people in

Figure 26. A storefront display of Japanese hanko, which are contemporary seals that function similar to signatures. Each hanko here is a common name in Japan, and every citizen is required to carry one to authorize documents. Image by Angie Harms. CC-BY-ND 2.0.

certain lines of business are required to carry three different types at any time.[18]

A long lineage of authentication was born out of the wait times and geographies of communication. Those wait times and geographies endure, and we continue to create modes of linking our bodies to our messages. The uncertain networks that were the intervening pathways for messages helped create the need for sealing practices, and those seals, as a result, provided a catalyst for a literate society. Wait times make us communicate differently, as we seek to create messages adapted to uncertain networks and the possibility of interception.

7 FIRST MESSAGES

The Los Angeles suburb where I grew up was designed like many suburbs in America and around the world, with an idealistic post–World War II faith that kids would be able to shout over backyard fences to their friends to meet up, run around the neighborhood, skin knees on sidewalks, all within the relatively safe confines of "the neighborhood." The neighborhood created its own community, visions of which trace back to the first indigenous communities on North American soil. Everyone was within earshot of one another, spoken languages were shared, and ideas were exchanged through face-to-face conversations. Handshakes, the exchanging of gifts, and even state or county laws were all signs that people had come together in the same place, talked with one another, and shared ideas.

Conversation has been the way for people throughout history to build bonds with one another. It's how they have fallen in love, passed history down from generation to generation, and made plans for the future. As psychologist and technology scholar Sherry Turkle writes, "Face-to-face conversation is the most human—and humanizing—thing we do. Fully present to another, we learn to listen. It's where we develop the capacity for empathy. It's where we experience the joy of being heard, of being understood."[1]

But conversations are limited in one important aspect. Whether it's an expression of love, a story about the past, or a group decision to pass a new law, a face-to-face message can be received only by someone who is within earshot. Conversations are good for sharing ideas among a small community, but they aren't well suited for sharing ideas with people or other communities who are located at a distance.

To deal with this challenge, we send messages. Messages bridge the distances between people. They allow ideas to cross borders and span cultures and communities. They allow for co-ordination across vast geographic expanses. Messages have been essential to human identity as we have communicated, collaborated, and cross-pollinated ideas across cultures and their landscapes. Often, messages have kept us intimately connected with those who have had to leave us to go out and work, fight, or explore in remote locations. Sending these messages, which have been essential to our relationships, requires us to wait for a response. Since we keep in touch with distant friends, lovers, and relatives through messages, waiting for word becomes an essential part of our lives as social creatures.

When did we decide that waiting for a response was worth our time? When did face-to-face exchanges cease to be sufficient for our needs? When did we begin to develop technologies for sending messages to distant people in lands unknown? When did humans first start sending messages?

The answer is a surprising one in that it far precedes written language and has been largely ignored by scholars. At least twelve thousand years ago—and as far back as forty thousand to fifty-five thousand years ago—as Aboriginal people migrated across the Australian continent, they did so with a sophisticated set of communication practices that allowed them to send messages across vast distances. While there was no alphabetic written language in these cultures, Aboriginal groups exchanged messages using message sticks. These wooden sticks looked a bit like a runner's baton, about eight to twelve inches in length and very slender. The leader of an Aboriginal group would dictate his message to a scribe, who would then carve the message out as pictographs or a series of cuts on the stick. Once completed, the message stick would be handed over to a runner, often the person in the group

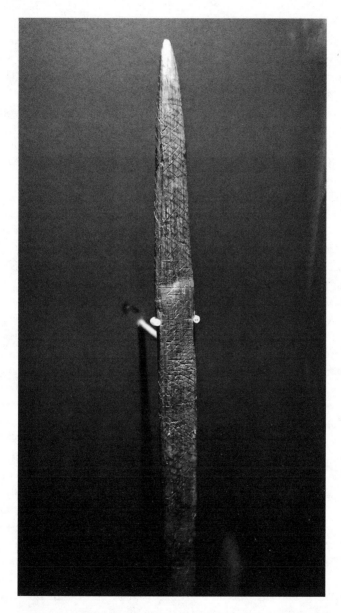

Figure 27. An Aboriginal message stick collected from Lake Condah, Victoria, Australia, 1893, from the collection of the Melbourne Museum. The European who collected this message stick wrote that the stick represented the messenger's credentials from the tribal chief. Image © 2016 Jason Farman.

with the most speed, endurance, and reliability. Taking his message with him both in memory as an oral message and carved onto the surface of the wooden message stick, the messenger would travel by foot or by boat to the recipient of the message. Message sticks became mankind's first "mobile medium."

The first Aboriginal groups to arrive on the continent of Australia might have used these messages sticks.[2] Dave Johnston, an Aboriginal archaeologist and founding chairman of the Australian Indigenous Archaeologists Association, noted that the complicated nature of inhabiting a harsh land like the continent of Australia probably required a sophisticated means of communicating across the land. "People have survived here for around fifty-five thousand years across climatic traumas. This doesn't just happen without extensive networks and movement along with other groups and their knowledge of food and resources," Dave told me. "This is the kind of social, ceremonial, and physical adaptation that can only happen with a sophisticated communication network."[3] So while oral histories describe the use of message sticks twelve thousand years ago, these tribes probably used this medium much earlier to communicate as they explored the land and shared knowledge with one another. What we know about message sticks comes mostly from a few surviving artifacts, the history passed down from generation to generation among Aboriginal people (who still consider these tools important to Aboriginal identity and culture), and from the writings of nineteenth-century anthropologists. As Europeans began to colonize the continent, anthropologists such as R. H. Mathews mentioned message sticks in their research, and in 1897 Mathews even provided drawings of the sticks. Message sticks weren't meant to last. They were made of wood, which deteriorates, leaving no archeological record. They were typically single-use, single-event tools that didn't contain an "archive" of knowledge in the same way as medieval documents sealed for posterity.

Sometimes a message stick was presented to honor someone, but the medium was more often used to communicate details about a gathering. The message stick functioned more like a contemporary text message than like a long love letter or a record of an important event. In essence, the sticks were designed to

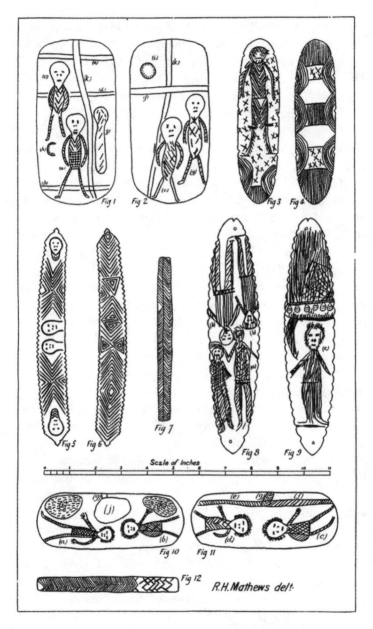

Figure 28. Drawings of various message sticks from European anthropologist R. H. Mathews, from his research in 1897.

disappear. Lynette Russell, a professor and the director of the Monash Indigenous Studies Center at Monash University in Melbourne, told me, "The very fact that they exist is remarkable. . . . These were probably never intended to have a permanency. They were meant to be an ephemeral thing that disappears like our emails. Somehow these are now here and curated."

I first heard about message sticks at a technology conference in Melbourne. Most events and conferences in Australia begin with a Welcome to Country ceremony led by Aboriginal elders from a local tribe. Bill Nicholson, a Wurundjeri elder, began the conference I was attending by noting that the site of the university was on tribal land, and that everyone at the conference had the responsibility to respect the space and understand the practices of the people who once lived there. Part of "Uncle" Bill Nicholson's presentation ("uncle" is a term of respect for tribal elders in Aboriginal culture) was about the use of message sticks in exchanging messages with other tribes. In addition to its communication function, the message stick served as a passport for the messenger through the lands of other Aboriginal groups, giving him unobstructed movement through these spaces.

It struck me that these sticks could be thought of as one of humanity's first mobile technologies; a device carried on a person and used for sending a message to a distant place. Throughout most of my work on the history of mobile technologies—and most other studies of the history of communication—I pointed to early "mobile" media like clay tablets and papyrus as the first forms of this kind of technology.[4] But papyrus dates from around 2900 B.C.; Aboriginal message sticks predate papyrus by as much as forty thousand, even fifty thousand years.[5]

In contrast to many of these early mobile media, a very limited archive of message sticks exists, and colonizers are largely responsible for those that have been preserved. Message sticks were, from the perspective of European explorers and anthropologists, windows into the practices of humans before we evolved into a technologically advanced and sophisticated species. This European perspective was itself extraordinarily simpleminded, overlooking how advanced Aboriginal culture was: Aboriginal groups had invented long-distance communication tools tens of thousands of years before the colonizers.

When Europeans settled in Melbourne in 1835, they believed that they had come across a prehistoric people who had been cut off from the modern world. Because the Aboriginal people lacked alphabetic written language and any obvious system for telling time, many colonizers believed that they didn't experience time. The Europeans believed that time was an abstract concept that was outside the Aboriginal realm of understanding. The indigenous people were seen as outside of time and lacking the technological know-how to tell time. And if Aboriginal people did not have a concept of time, the Europeans assumed that meant that this group of "prehistoric" people did not wait in the ways that advanced cultures do.

Since technology and time refine and amplify each other—that is, as a technology divides time in more accurate ways, we shift our experience of time to match the standards of that new timekeeping technology—it is logical that cultures with different modes of timekeeping technology would wait in different ways. Different expectations would govern what constituted being late or keeping someone waiting.

But the European colonizers failed to recognize message sticks as a form of timekeeping technology, and this short-sightedness shaped inaccurate biases about Aboriginal people. The examples that I have discussed throughout this book point to the ways that time and messages are deeply connected to geography. As a message travels across the land, whether on foot or on horseback, by train or by telegraph wires, the speed of a message is directly connected to how quickly it can traverse the landscape. When an Aboriginal group sent a message stick to another group, the timing was of key importance. Many of these message sticks were invitations to another tribe to join the senders for a gathering called a corroboree. The message stick and the accompanying oral message specified a particular location and a particular time for the gathering. The timing of the corroboree would have to take into consideration the time it took for the messenger to get to his destination, deliver the message, and return with word that the other tribe had accepted the invitation, as well as the time it would take for both tribes to travel to the common location. Timing was thus embedded in the exchange of these message sticks.

The distorted biases held by the colonizers made their way into nearly every anthropological account of early contact with Aboriginal groups. They were so pervasive that they ultimately shaped stereotypes about Aboriginal people that are held to this day. Contemporary scientific studies of such issues as mental health in Aboriginal groups include discussions of how indigenous people's experiences of time tends to focus on the immediate and not the long term. Aboriginal people are even described as remaining "timeless" rather than being constrained by any modern relationship to time and timekeeping.

These persistent stereotypes even made their way into discussions about my meetings with the Wurundjeri people. I had set meetings with them to learn about the place of message sticks in their oral histories, but many white Australians cautioned me about trying to impose my own notions of time onto these cultures. With this in mind, I was hyperaware of my own preconceptions about time while approaching these meetings. I wanted to respect another culture's relationship with time, yet I didn't want to do so with a lens that brought long-held European biases about a "primitive" culture and its lack of advanced timekeeping practices.

I was aware of cultural differences as well as biases when I set up a meeting with one of the Wurundjeri elders, Aunty Joy Murphy Wandin. I had contacted her in the weeks before my flight from Baltimore to Melbourne, and she planned a meeting for us near the tribal land of Coranderrk, some thirty-five miles northeast of Melbourne. The day after I arrived, I got on the train to Lilydale station at 8:45 in the morning, still jet-lagged from my twenty-two-hour flight. The train car was full, and the day was bright and sunny, the weather almost perfectly matching the temperature back in Maryland, just swapping out the spring buds on Washington's cherry trees for yellow leaves falling along Melbourne's laneways. The seats were filled with students, businesspeople, and

workers in transit along a railway that once threatened the Coranderrk "reserve" for Melbourne's aboriginal tribes not too long
after Europeans arrived in Port Phillip.

During the journey, as I was on my way to learn about the first
mobile medium humans used, I kept an eye out for the various
mobile media that we use in our own era. On a seat across from
me, a woman was on page twenty-five of a massive book she had
checked out from the public library. (I notice from the cover that
it's a trilogy collected into a single volume, easily spanning more
than eight hundred pages and definitely not the most mobile of
books I've seen.) A student in a maroon sweater and button-down
shirt held his phone twelve inches from his face, reading the news.
Two other students sat next to each other, each wearing headphones.
They were obviously not together on this journey, as she pressed
her cheek up against the window and he sat halfway out of his seat,
his high socks and running shoes stretching into the aisle of the
train. He tapped his foot to music while scrolling through social
media. The train was silent except for one woman doing business
on her phone, talking loudly about a university student needing
help with a scholarship. A businessman read through a brief
printed out on a thin stack of papers stapled together. The loud
woman on the phone, like most of the car's passengers, got off at
Glenferri Station, which serves Melbourne University of Technology. The train pulled away again. Street art and graffiti filled most
blank surfaces between train stations. Modern high-rise apartment
buildings sat next to homes with ornate Victorian trim. Eucalyptus
trees grew right up against the fences bordering the train tracks.
We raced past a platform where about twenty children and teens
in school uniforms waited for their train to school. Not a single
child was on a phone; these children either chatted with one
another or stared off, bored. Around this group of school-age
kids, nearly every adult waiting for the train stared at some kind
of device, headphones plugged in. Soon I was almost alone in
my train car, accompanied only by an elderly man pushing his
two-year-old grandson in a stroller. The man took out his phone
and captured a picture of the two of them out on their morning
adventure. Mobile technology was ubiquitous—except for the

schoolchildren, who I presumed were banned from bringing electronic devices to school—and were used to communicate or pass
the time as people waited during their transit.

I arrived early. I exited the station and looked for Aunty Joy.
The station was small and quaint. She wasn't there. When I
checked my phone, there was an email from her letting me know
that she had gotten the time wrong. She was some distance away,
running errands, and would be an hour late. She suggested that
while I wait, I run across the street to get a coffee and enjoy the
sunshine on this beautiful day. I quickly responded that the delay
was no problem, and that I'd be back at the station to meet her at
11 A.M. I sat in a coffee shop–*cum*–ice cream shop, drinking a latte
from a ceramic mug, and read quotations painted on the wall
about the history of ice cream (most of them referencing moments
that had taken place in Baltimore or Washington, places I've called
home for many years).

An hour later, I stood in front of the small train station again
with the sun on my back. The wind was cold and, typical of Melbourne weather, the clouds occasionally blew in and covered up
the sun. I rocked back and forth on my feet to keep the blood
flowing. A car pulled up with two people in the front. I pulled my
bag up around my shoulders, ready to meet my hosts for the day.
The passenger in the car got out after kissing the driver good-bye.
She slung on her backpack and entered the station. The car pulled
away. This scene was repeated a dozen times before I stopped
anticipating that the vehicle was for me. I sent another email to
Aunty Joy to let her know I was out in front of the station in a
dark blue sweater.

An hour later, I laughed to myself that I was writing a book
about waiting and was left waiting while trying to do research for
the book. I reflected on the fact that lateness is culturally specific
and differs in scope from society to society. I noticed the buses
nearby departing for Healesville, a town that borders Coranderrk.
On my phone, I wrote another email to Aunty Joy, telling her that
I was going to eat lunch at a local fish and chips shop and asking
whether I should take the bus to Coranderrk. By this point, I had
checked my email obsessively every five minutes to see whether

she had written. I hadn't heard from her in two hours. I tried to maintain some self-control over how often I checked for a message, worried that my phone's battery would die and I'd be left out in a part of Australia without a map or a sense of where to go.

After lunch—a greasy, mouth-burning affair—I went back to the train station for the third time. I again stood in the sun and mapped out how long it would take to walk from the Healesville bus stop to Coranderrk; it looked like twenty-minute trek. At least I could go and see the space, chat with a few people who were there, and check to see whether Aunty Joy was all right. I got on the bus, shooting Aunty Joy yet another email to let her know that I was on my way via bus and would be at Coranderrk around 2 P.M.

The landscapes were amazing. Hills covered in vineyards. Distant mountains. Cattle grazing everywhere. These were the properties that settlers had occupied as they began colonizing this part of Australia. Initially, European colonizers pushed the indigenous people onto these lands outside of the city of Melbourne. But once the Aboriginal people cultivated the lands, the European settlers saw the value in these properties and kicked them off the land and relocated them to a more distant space. This pattern was repeated over and over again. Land was snatched up at a rapid pace, eventually encroaching on the Coranderrk land set aside for Aboriginal people. Coranderrk itself became subject to this same cycle; one of the key arguments for the land being given back to white settlers was that it could be sold for a hefty sum in order to finance the railroad to Lilydale, the same train that I had taken earlier in the day.

I got off the bus and pulled out my phone, mapping the walk to Coranderrk. The bus's only stop at Healesville left me with an hour's walk. I asked the bus driver about a bus stop closer to Coranderrk. "Oh yeah, that's back that way. I don't stop there." When I told him that I was going to see Coranderrk (which resulted in a blank stare; he'd obviously never heard of the place) and that the map showed that I had an hour walk ahead of me, he responded, "Oh, that shouldn't take you an hour!"

An hour later, after wandering along highways with no sidewalks, I arrived at Coranderrk. The property was fenced off, with an overhang of chain link to keep people out. I stood at the entrance, which was a large metal gate closed and padlocked shut. There wasn't a person in sight. I shot off yet another email to Aunty Joy telling her that I had arrived, that the gate was locked shut. Was there any way in?

In my mind, as I waited for Aunty Joy, I continually thought about cultural differences involving time. I thought about the stereotypes that are applied to marginalized groups and their seeming indifference to time or constant lateness. Back home in the United States, the month before my visit in Melbourne, people had criticized presidential candidate Hillary Clinton for making a distasteful joke about "colored people's time," a stereotype casting African Americans as habitually late. When used outside of the African-American community, the expression functions as a marginalizing statement that positions people of color as "clock challenged," as Karen Grigsby Bates says.[6] These aspersions constitute an exercise of power, an implication that certain groups are not aligned with the dominant forces of time and the connection in the dominant culture between time and value ("time is money").

I then received an email from Aunty Joy. When she stopped at home before our meeting, she had run into her house and left her smartphone in her locked car. When she came back outside and was about to depart, her car key had broken off in the lock. She had been unable to access her main mode of communication, her mobile device. She had no spare car key and instead of traveling to meet me, she was obliged to wait for a roadside service company to arrive and unlock her car door.

I received this message as I stood in front of Coranderrk, gates locked and no one there to show me the property. As I wandered around Melbourne, as I waited to coordinate with Aunty Joy during my journey, the relationship between the experience of time and the experience of a landscape stood out to me. The two seemed intimately bound together. Journeying, waiting, and the duration of time are linked. The notion that a culture, especially one as tied to its landscapes as Aboriginal groups were, could be without time was absurd. Duration has distinct paces and tempos for different cultures and people in different parts of the world, but the journeys that were so important to Aboriginal people were indelibly linked to time, and they still are.

Standing in front of Coranderrk was key for me to understand the relationship between the messages that had once been sent from this location to the newly established Parliament in Melbourne, advocating for the continued support for the people on this land. The people who lived here in the mid- to late 1800s were the last group of Aboriginal people to send message sticks. As they lived and worked on this land, they began adopting written letters in English as a means of communicating with the colonizers. Within a few decades, message sticks almost entirely vanished, effectively condemned to forced obsolescence by the government, which instituted a new law that separated families and imposed assimilation on young Aborigines.

At the center of this story, I found William Barak. Barak was an ancestor to Aunty Joy, from whom I would later learn about his life. Though the timing never worked out for us to meet face-to-face, I did eventually get to hear Aunty Joy's stories about Coranderrk and her Great-Great-Uncle William Barak, the last tribal leader of Coranderrk. Through messages we exchanged, she shared her knowledge of the Wurundjeri people and their use of messages throughout history.

Joseph Wandin, a relative of Aunty Joy from several generations earlier, finished writing down all the words dictated by the leader of the Aboriginal group, William Barak. On this day in 1881, Barak was the group's *ngurungaeta,* or the leader of this tribe. Wandin sealed the letter and, along with all of the tribal elders, began walking toward Melbourne. The walk from Coranderrk, the land they had been placed on by the colonizers, typically took a little more than twelve hours. This deputation of tribal elders walked the long journey wearing their best suits. Coranderrk was in a remote region, far from the waters of Melbourne's Prince Phillip Bay. It was lush land at the foot of mountains among rolling hills and eucalyptus trees. Embarking on the long journey was a group of Aboriginal leaders with the goal of delivering a single handwritten letter to Australia's Parliament, advocating the group's independence from

Figure 29. Sixteen Aboriginal men, all wearing European-style clothing at Coranderrk, with William Barak seated in the center, back. This photograph was taken toward the end of the nineteenth century. Collection of the State Library of Victoria. Used by permission of the Wurundjeri tribe.

the European government's "protection board." The protection board did everything but protect Aboriginal people; in fact, the Board for the Protection of Aborigines is the nemesis of a story in which William Barak is cast as the hero.

Barak had recently broken his leg, and it had not yet fully mended. As he struggled down the pathways through rolling brush and thickly forested regions, he led the group of twenty-two emissaries from Coranderrk. He set the slow pace across the land, and the journey took them more than a day. The delegation's walk connected them to the practices of their ancestors, who had similarly walked this land. To make a long walk to deliver a message was part of Aboriginal practices since these cultures emerged onto the continent of Australia. Here, instead of a message stick accompanied with an oral account of its message, Barak and the other members of the tribe were carrying a letter written in English. The act of delivering a message was something that resonated across both cultures, so Barak was prepared and skilled in the ways of using messages as a way to bridge the differences between his people and the colonizers who had taken over the land when he was a young boy.

This approach was emblematic of Barak's outlook on the future of Aboriginal life: In order to survive, they must find compromise and common ground. Using modes of diplomacy familiar to the Europeans, he would successfully advocate for his people. For many years, his people had thrived at Coranderrk. As Barak made this painful journey to Melbourne, all he had worked for to keep his people safe and thriving in the face of colonialism was in jeopardy. They were about to lose it all after already losing so much after the Europeans had arrived on their shores.

Barak's injury meant that the journey to deliver the message was slow. The group would have to wait a bit longer to get its message heard. Yet the slow pace was significant because Barak's presence as the tribal leader was a sign that this was an important message. Messages were connected to the land for Aboriginal tribes; stories, information, and history all were linked to the places they shared and the journeys they took to arrive. When the medium was the message stick, the journey typically had been made not by the tribal leader but by a messenger sent on behalf of the *ngurungaeta*. As with a medieval seal, the message stick might contain elements of the body of the person sending the message. Some early colonizers collected message sticks that contained locks of hair from the messenger as a means of authentication.[7] The messenger would use this embodied mark as proof of the authority of the message and its "author." Speed of delivery was important, which was one reason why an emissary was sent instead of the tribal leader. In this instance, however, Barak and his delegation would slowly travel throughout an entire day and night to Melbourne, making sure that the letter and its author were united in presentation. Barak requested that the government keep his people on the Coranderrk land that they had spent years cultivating into a successful hops farm. They were self-sustaining and asked only to be self-governing.

Barak was born in the 1820s and witnessed the first arrival of Europeans in Melbourne. At the time, his older cousin, Wonga, was the *ngurungaeta* of the Wurundjeri. As colonizers began arriving, the population of Aboriginal people throughout Australia began to be decimated. The remaining indigenous people were isolated on reserves, often in terrible conditions. Wonga and Barak, having been displaced already, had selected the site of Coranderrk for the site of the Kulin clan, a group of tribes united by language and homelands near modern-day Melbourne.

"By his mid-30s, Barak had seen the birth and growth of the city of Melbourne; the surge of half a million newcomers in search of gold, the equally rapid reduction of the Kulin population, and the containment of the survivors on a handful of missions and government reserves. But Barak was not just a spectator—he also became an active participant and an agent of change," argue Giordano Nanni and Andrea James. While writing a play about the history of Barak and Coranderrk, they did extensive research on the history of the site, both in the archives and with oral histories of Coranderrk descendants. They noted how Barak "transformed a place of intended incarceration into a vibrant and empowered community."[8]

Yet the Board of Protection continually threatened to remove them, citing such false concerns as "winters . . . too cold for these people to be living in this remote land." The people at Coranderrk had worked closely with a European pastor, John Green, who had been appointed general inspector of the Aboriginal stations around Melbourne. Green believed that the Aboriginal people should be self-governing and self-sustaining. For his views, Green was eventually terminated from his post; a series of successors followed, each bringing his own ignorance of Aboriginal culture and his own abuses to the people at Coranderrk.

Coranderrk was a thriving, self-sustaining farm, producing the best hops in the region. Its success made other local farmers, all of European descent, perceive it as a competitor. These local farmers criticized the Board of Protection for putting money into Coranderrk, thus financing their competitor. During the 1870s, Parliament was under increasing pressure to finance such infrastructure projects as expanding the railroad lines, in order to bring Melbourne into the modern age. The sale of the fertile land of Coranderrk and its profitable hops farm was seen as an easy way to subsidize these projects and bring Melbourne the modes of transportation that would help it come up to speed with the rest of Western society.

Everyone living at Coranderrk hoped that it would be their lasting place, where they would be protected from the encroaching European settlers in a "homeland where they could avoid the deprivations of disease and frontier conflict."[9] To protect this land and advocate for his people, Barak engaged in a "paper war," as aboriginal historian Diane Barwick terms it.[10] Barak and tribal elders at Coranderrk translated the practices of the message stick into a letter-writing campaign to save Coranderrk. Barak would dictate a message to be inscribed by one of the young members of the tribe who had learned English, such as Joseph Wandin or Thomas Dunolly. Delegations of elders would walk the twelve-hour distance to hand-deliver these letters—a process they repeated several times throughout the history of Coranderrk.

As befell intercepted letters discussed in the previous chapter, many of these letters were hijacked along the way to the recipients in Parliament and never delivered. As Nanni and James write, "The Board, unable to either silence or ignore these protests, responded by intercepting communications, threatening letter writers and members of deputations, and sought to dismiss their petitions by disputing their authenticity."[11]

These impediments to delivery highlight a difference between message sticks and letters. As we have seen, the message stick gave the messenger free passage through hostile lands. The neutrality of a messenger was not to be disturbed. Properly authenticated messages (bearing strands of the sender's hair, for example) were considered of utmost importance in the exchanges between Aboriginal tribes, so messengers and their communications were protected. Cultural practice ensured that messages got through without delay. This is an approach that carries over into the position of contemporary net-neutrality advocates, who believe that the messages and data sent across fiber optic lines should not be purposefully impeded by any entity.

The delegation carrying letters to Parliament, like earlier messengers carrying the sticks across the land, were go-betweens empowered to bridge cultures. The letters written by the tribe at Coranderrk were used to translate between European and Aboriginal culture, as message sticks had bridged Aboriginal cultures. Before colonizers arrived, rich cultural differences prevailed among the more than five hundred Aboriginal tribes or language groups that inhabited the continent. Though the idea of the "Aborigine" allowed colonizers to homogenize a truly complex array of cultures under a single heading, when these groups interacted with one another, ideas, resources, cultural traditions, and belief systems were exchanged.[12] The Aboriginal people comprised diverse groups that used messages as a means of cross-cultural exchange.

Messages and messengers are thus bridges between cultures, and, as when explorers are sent out to report back new knowledge about what they find, people wait for their return to expand the scope of knowledge. Like the bridges that connected people during the American Civil War, messengers are go-betweens that connect people and establish coherent societies. The messenger, as a go-between, is someone who can translate across cultures to point out the overlaps and relationships between these societies.[13] These go-betweens also shape the sense of place of these tribal groups. As messages go out and take a certain length of time to travel, the duration of this journey reiterates the geographic location of the place a tribe calls home. Home is defined in relationship to that which surrounds it. The Wurundjeri, across the ages, note that home stretches from present-day Melbourne "across the mountains of The Great Dividing Range, west to the Werribee River, south to Mordialloc Creek, and east to Mount Baw Baw."[14]

Barak approached his use of letters through his understanding of message sticks. Letters were tools of connection that translated between cultures. He was ultimately using the colonizers' tools of alphabetic written language, pen, and paper to fight the impact of colonization on his people. He was the go-between who translated for the Europeans what displacement would mean for the already-traumatized people at Coranderrk. In his letter, he urged Chief Secretary Graham Berry to dissolve the Board of Protection and allow Coranderrk to be self-managed by the Aboriginal people.

Barak's messages made him the catalyst for an investigation to establish whether the Board of Protection was attempting to push the Aboriginal people off the land for profit. As a result of the inquiry, Chief Secretary Berry ordered that Coranderrk be permanently reserved as a "site for the use of the Aborigines."

Coranderrk represented a major shift in Aboriginal life. The advocacy for this land symbolized a shift from traditional modes of communication using message sticks to adopted ones, using letters written in English. The young Aboriginal people who wrote these letters would also be caught in the shift. Though the Board of Protection was defeated regarding the land use of Coranderrk, it instituted a new law in 1886, the Half-Caste Act, which required any Aboriginal person under the age of thirty-five who was not full-blooded Aborigine to leave the reserves and stations like Coranderrk. These young people were subsequently forced to assimilate into the society established by the European colonizers. The bulk of young people who farmed the land of Coranderrk had to leave. Families were split up, and most of the population remaining in Coranderrk were elderly people gathered from various tribes in the region. Coranderrk began to decline and was eventually shuttered in 1924. The land was given out to farmers and used, at one point, as a settlement site for soldiers returning home from World War II—though Aboriginal soldiers were excluded. Eventually, Aunty Joy Murphy Wandin's family purchased the land from the new owners, reestablishing Coranderrk as a site for descendants of the original Coranderrk families to gather and maintain their history.

When I eventually connected with Aunty Joy Murphy Wandin on the phone and via email to learn all I could about the history of Coranderrk and her relative William Barak, I asked her whether any of the Wurundjeri use message sticks any more. "Oh no," she replied, "the Half-Caste Act wiped out much of our traditions by forcing the young people to assimilate." She did note, however, that message sticks are still highly regarded in the culture as a link to past practices. Many Aboriginal artists create contemporary message sticks as a way of maintaining the device as a vibrant piece of their history and culture. Lynette Russell echoes this sentiment, "We can't necessarily read them, but we are inspired by them. By continuing on in the tradition of making message sticks, we have a way of creating the art of our ancestors in our own time."

Many factors are at play when technologies go obsolete and are replaced. In the case of Aboriginal peoples and colonizing Europeans, the dominant culture imposed its own practices, which its people assumed to be more technologically advanced. This ultimately erased message sticks as a practical medium. Our own time offers many parallels, as innovations erase practices that are considered slower or out of date. We live in an era in which technologies become obsolete at a head-spinning rate, many in service of making communication connections faster and making us wait less. Because the newer technologies are faster, the obsolescence of what came before is sold to us as a benefit, not a loss. We can trace this from the latest iPhone launch to the abandonment of the pneumatic tube mail system to the loss of the Aboriginal message stick. Walter Benjamin described this trend in his essay "Theses on the Philosophy of History." Benjamin describes the "Angel of History," who looks backward at the wake of human events, all under the umbrella of innovation. What the Angel of History sees, however, is not "a chain of events"; instead, "he sees one single catastrophe which keeps piling wreckage upon wreckage." The storm that creates this chaos is called progress.[15]

Under the auspices of progress, many of the messaging technologies discussed in this book have been made obsolete. Instantaneous culture creates a context where abandoning a past technology for a faster one is common sense. As I write this, technological obsolescence will continue and make even the contemporary examples in this book obsolete. In the coming years, we will clamber toward innovation in the hope of finding technologies that will connect us at ever-accelerating paces. Those technologies, too, will go obsolete in favor of new ones that promise higher speed and more reliable connections.

In contrast to this scene are the people who wait and embrace waiting. These are the maintainers; the people who choose a pace that may not fit with the constant acceleration promised by new technologies.[16] The maintainers are defined by their interest in finding links between our media across eras and allowing us access to them. By maintaining what exists, we preserve certain modes of time in a culture as they are linked to the technologies of the past. By looking toward the promises of maintenance instead of the promises of innovation, perhaps we can all wait differently and wait a little better.

8 TACTICS FOR WAITING

Herbert Blau and the actors he was directing were nervous. Standing backstage before their performance of Samuel Beckett's *Waiting for Godot,* they peeked through the curtain at the audience that had gathered. The venue hadn't hosted a performance in forty years, but each seat was occupied. The play, which premiered four years earlier, in 1953, was being performed for a unique audience: prisoners in San Quentin State Prison, just north of San Francisco.

Blau had chosen the play in part because it didn't require female actors, and at the time women weren't allowed in the prison. The warden warned the performance troupe, the Actor's Workshop of San Francisco: "If these guys don't like you, they're going to let you know it."[1] Blau was understandably uncertain. After all, this is a play in which nothing happens. The two main characters, Vladimir and Estragon, wait for the arrival of Godot; Godot never arrives. The main characters aren't even sure whether they have the right day for the appointment. The stage design was stark, containing only a single leafless tree. Before the performance, Blau asked the prisoners to think of the play as a bit like jazz, to listen and experience whatever connected with them.

The curtain was pulled across the two boxing rings that had been pushed together to form the stage. Besides standing in a prison for the first time in their lives, about to perform a complex play for an audience of convicts, the actors had other reasons to be nervous. When *Waiting for Godot* opened in Paris, the audience and the critics hated it. It was shut down before it completed its run. At one point in the play, Vladimir and Estragon speculate that suicide may be the best answer to the suffering created by their endless wait, and Estragon asks Vladimir whether he has some rope. At the London premiere, an audience member responded, "For the love of God, somebody get them some!"[2] At the same London premiere, when Estragon described the characters' situation—"Nothing happens, nobody comes, nobody goes, it's awful!"—an audience member replied, "Hear! Hear!" The audience also felt the pain of waiting, hoping the play would end quickly.

But five minutes after the curtains were opened at the San Quentin performance, something fascinating happened. The twelve hundred prisoners in attendance fell silent. Eugene Roche, the actor playing Vladimir, said he had never performed in front of an audience so attentive before. "It was dead, dead silence. Now you usually hear silence but you hear the rattling of programs, people adjusting their seats, coughing, and so on. None of that."[3] This audience knew what it meant to wait for what never arrives. "Godot is society," one prisoner reflected. Another said, "He's the outside." One of the teachers in the prison said, "They know what is meant by waiting . . . and they knew if Godot finally came, he would only be a disappointment."[4]

Beckett's play, in its many violations of theatrical norms, strips away plot expectations to make a comment on the human condition. Godot symbolizes whatever we wait for, whatever we long for, whatever we rely on to save us from our current state of uncertainty and despair. Godot represents the promise of what might come on the other side of our waiting.

After a performance of *Waiting for Godot* I saw in Los Angeles, I reflected on all the ways that I had waited, from eagerly waiting for the day of the performance to arrive to waiting in traffic to get to the theater. I waited in line to get inside the theater. I waited in anticipation for the curtain to rise. I waited during the performance for action to happen. But the only action that happens in the play is more waiting. I waited for something that never arrived, and the play asks us to reflect on this as our lot in life. Meanwhile, as we wait, we find ways to pass the time just like Vladimir and Estragon: We interact as social beings, complain, discuss food, tell stories.

Though reviled by its first audiences in Paris and London, the play has gone on to connect with broad swaths of people—from prisoners in San Quentin to new audiences in a digital age—because it shows how time flows through us and changes us. Day after day, as we wait for the things we desire, we become different people. In the act of waiting, we become who we are. Waiting points to our desires and hopes for the future; and while that future may never arrive and our hopes may never be fulfilled, the act of reflecting on waiting teaches us about ourselves. The meaning of life isn't deferred until that thing we hope for arrives; instead, in the moment of waiting, meaning is located in our ability to recognize the ways that such hopes define us.

Yet despite the ways that waiting can be instructive and meaningful, it is still regarded as punishment and a pain to endure. For the prisoners in San Quentin, *Waiting for Godot* highlighted the many ways that make waiting, isolated from society, one of the most excruciating forms of punishment. Waiting is made visible as the price to be paid for crime. As countries around the world exact punishment for crimes, they enforce on convicts an experience of time passing as justice and, criminologists hope, a means of rehabilitation. Time is meant to be slow and thick and noticeable for these inmates.

The rest of us, in the small slivers of time we have to wait, may feel as if we are given a glimpse of how these prisoners feel each day: powerless, punished, in someone else's control. Unlike time as it is experienced when we are being productive or enjoying an activity, waiting is noticed and lived. Waiting becomes obvious. We feel uncomfortable, uncertain, anxious.

Though we notice waiting in these moments of endurance and pain, I believe that we don't really understand it. Waiting tends to hide the reasons it exists. As we go about our day, time can seem transparent, not drawing explicit attention to the work it does on our lives. Media and time are noticeable or invisible based on the relationship they bear to content. In some contexts, we are meant to notice the medium of an experience; in other contexts, it is designed to disappear. When we attend a movie, the lights go down and we are meant to simply connect with the content of the film; we notice the medium only when it breaks down, such as when the volume is too soft or there is a glitch with the projector. Other media are meant to be noticed as tools for communicating the message; think of cubism, Beckett's minimalist style of theater, a contemporary artist who uses an obsolete medium like Polaroid film, or musicians who release an album only on vinyl. Time, as a medium, works in similar ways. It can be conspicuous, noticeable to a lesser degree, or invisible, seamlessly enabling us to accomplish our goals. Time tends to be invisible when we feel that we are making the most of it.

Waiting, in contrast, draws attention to itself, as a medium that we can't ignore as we attempt to accomplish our day's tasks. Having to wait feels like a direct affront to the goals we may have as we work, or when we hope to engage in leisure activities. Waiting is a hurdle between us and our plans. We see wait times as burdens and obstacles, but we rarely notice the larger reasons for this perception. We don't truly see the larger structural reasons why waiting is so reviled. Why is waiting seen as the ultimate punishment and the hurdle to living the good life? How can we reconcile these beliefs with the fact that waiting is time's great teacher about who we are and who we hope to become?

As I've worked on this book, many people have asked me what "waiting differently" might look like. We can't really avoid waiting, they acknowledge, so how do we deal with it in efficient ways that minimize its impact on our goals? How can waiting be productive rather than a waste and a burden?

I think these questions reinforce a stereotype. These questions do not acknowledge that waiting can be a beneficial activity. These questions assume that productive time, in which we pack each minute full of tasks that meet our to-do list, is the best path toward living a full life. As I hope to have demonstrated in this book, there are times when waiting can be essential to certain modes of learning, creativity, and connecting.

The questions I have fielded reflect value judgments about how time is and should be used: Productive time is efficient, while waiting is inefficient; productive time seamlessly weaves into our lives, while waiting is full of seams and puts our lives on pause. Productive time is invisible and waiting is conspicuous. Time's movement between invisibility/visibility, efficiency/inefficiency, and seamless/ stitched mirrors the ways that designers have described our encounters with emerging technologies. Mark Weiser, the computer scientist and designer who coined the term "ubiquitous computing" to describe the proliferation of media that seamlessly "weave themselves into the fabric of everyday life until they are indistinguishable from it," noted that these kinds of media should be "literally visible, effectively invisible."[5] In other words, we can see them if we look, but as we go about our day-to-day lives, these media should blend in so well that we don't notice them. Time is meant to parallel this way of blending in with our lives, goals, and actions. We can see it and notice it, but it is "effectively invisible" when we're going about our days. Media and time should mesh seamlessly with our lives. When the seams reveal themselves, when our media break down or when we are forced to wait, these experiences run counter to the ideal mode of efficient and seamless interactions.

Looking at the seams, however, allows us to see how things are put together.[6] Waiting forces time to be visible, as loosely stitched and seemingly inefficient. We are able to see the cultural assumptions around what our society assumes to be good or wasteful uses of time. As we begin noticing these times, we can see the work they do in a society, both good and bad. By making us students of the seams, wait times can teach us about the forces in life that have shaped our assumptions about time, efficiency, and productivity. Once we study these things in our lives, we can begin developing tactics for dealing with our wait times.

"Tactics" is a term drawn from military usage. Strategies are plans of action directing a military force when attacking another, and tactics are responses to conditions on the ground.[7] In this vein, time is imposed on us by our cultures, by the technologies that have regimented time down to the nanosecond, and by its own finite nature and the fact that we're going to live only so long. In response, we must develop tactics for dealing with time and waiting. These aren't tactics to eliminate waiting; instead, these are tactics for teaching us how to learn from the seams. These tactics have the potential to reorient us in profound ways, transforming our perspectives on our wait times. Such renewed perspectives transform waiting from a burden to a springboard toward things like creativity, social critique, or reflection on our inner state and the state of our relationships.

It is impossible to answer the question "What does it mean to wait in the digital age?" because waiting's meaning is always based on its unique context. But we can explore why we respond as we do to waiting when it appears. Wait times—in an age that values the instant—are instructive, but first we must allow ourselves to *experience* moments of waiting. If my undergraduate students are representative, collectively we are fleeing from wait times. Even in the most mundane and minuscule moments of waiting, we find some way to occupy ourselves. When my students are confronted with wait times (and the boredom that accompanies this waiting), they pull out their phones and check their text messages, social media, and email, or play a game. (Some students seem to want to report an activity for every moment of the day, so even if their self-reporting isn't completely accurate, the ambition to fill the entire day is telling.) If we want to be tactical about how to learn from and respond to our wait times, we must allow ourselves to acknowledge these moments.

The challenge of dwelling in our wait times is that our attention is constantly tugged at. As Nobel Prize–winning economist Herbert Simon has noted, societies that are surrounded by information-rich media typically face a scarcity of the resource that the information media consumes. "What information consumes is rather obvious," he writes; "it consumes the attention of its recipients." The consequence is "a poverty of attention" and our subsequent attempts to spread our attention efficiently across this overabundance of information.[8] The stakes of this scarcity are high. As legal scholar Tim Wu argues in *The Attention Merchants*, the ways we direct our attention don't just affect how well we use our time or the discussions that end up dominating national discourse; instead, "the very nature of our lives is at stake. For how we spend the brutally limited resource of our attention will determine those lives. . . . When we reach the end of our days, our life experience will equal what we have paid attention to, whether by choice or default."[9]

While I have focused on the moment when people wait for a response to a message—a moment that shapes our social identities and the ways we build knowledge of our world and universe—the number of ways that we wait is extraordinary. Paying attention to our wait times opens the possibility of learning from waiting and building on this knowledge as a force to positively shape our lives and our societies. How we wait comments on our society's notions of power, efficiency, and ambitions for the future. If we don't pay adequate attention to these wait times and the force they exact on daily life, then we will simply be pulled into a future of someone else's making rather than attempting to respond tactically to the strategies of those who impose their time upon us.

The first tactic in learning from our wait times is to shift away from a focus on the feelings of waiting (anger, boredom, discomfort, and longing, among others). It is as simple as asking, "Why am I waiting?" The initial answers will probably be shallow responses to a complex situation. You may be waiting because your friend is running late, or you may be waiting because your boss is wrapping up a previous meeting. You may be waiting because the train runs only every twenty minutes on weekends. You may be waiting because the restaurant is packed with people and the kitchen is slammed with orders. You may be waiting because you are incarcerated in San Quentin State Penitentiary until your parole hearing.

A follow-up question that helps us move from shallow to complex answers is "Who benefits from my waiting?" That is, while I may perceive waiting as imposed on me and often either inconveniencing me or denying me my ability to control my own time, it is important to ask who (or what) benefits or profits from my waiting in this moment. Sometimes we are the beneficiaries of our own wait times. Waiting can be an investment that pays out to us in a range of ways. Waiting can be a way for us to save or accrue money in a retirement account (rather than spending it when the first need arises), or it can be a way for us delay gratification. I might wait now to get something better on the other side of my waiting. We often attribute our ability to wait to the building of patience, an esteemed character trait. Those around me might benefit from my waiting, such as the drivers of other cars at an intersection where I am waiting at a traffic light. Yet waiting might also reveal structural benefits, when those in positions of power reiterate that power by making us wait. As discussed in chapter 3, the design of waiting rooms for some welfare offices in South America (and elsewhere in the world) make people wait in uncomfortable and uncertain ways in order to reaffirm the powerlessness of the poor in that society. Similarly, making you wait for a meeting might be a boss's way to confirm his or her position of power in that relationship. The same can be said of romantic relationships: The person who makes a partner wait is communicating a claim for power. Ultimately, making others wait is claiming priority of time. This observation can lead to another telling question: How am I making others wait? While some people in positions of power over us make us wait, are there people in our lives over whom we similarly exercise power by imposing waiting on them?

These examples reveal two important themes for developing tactics of waiting. First, waiting is a collective experience of time. Second, waiting is a way that power is exercised. This first point reorients us in potentially positive ways for how we encounter waiting. One of our reasons for feeling that waiting is a burden is that we believe our time is distinct from other people's time. In contemporary Western societies, we tend to value individual time over collective time. My time doesn't correspond with your time; we're each living in our own time, and you often get in the way of my using my time effectively. Instead, if we see time as collective rather than individual, we can see how our wait times can benefit those around us.

Recently I was in line at the grocery store with two shoppers in front of me and a long line of people behind me. I had a cart full of groceries, including many frozen items that were slowly thawing or melting as I waited for my turn to check out. The lines were all packed on this busy Saturday morning. The woman at the checkout counter, a young mother with her toddler sitting patiently in the cart, was paying for her groceries. She had to split the payment between cash and food stamps. She was also making sure that all her coupons were counted, because she believed some hadn't been tallied correctly. The store manager had to come over to answer the customer's questions about store specials and override the system to correct the errors. The customer in front of me, a well-dressed elderly woman, turned to me and rolled her eyes. As the process dragged on, the two men behind me voiced their annoyance that they were being kept waiting. The young woman ignored them. As she made sure that everything was correct, the men continued, "Oh, come on!" turning to other people in line to exchange facial expressions of frustration.

From the perspectives of some of these patrons, the woman who struggled and deliberated over how to pay for her groceries was making them wait. She was wasting their time. She was holding up the line and violating the social expectations of moving quickly at a peak time for grocery shopping.[10] Her inquiry about the coupons was perceived as a selfish act because it made others wait. However, by prioritizing individual time instead of collective time, the people in line behind the woman paying for her groceries were creating a split that fostered frustration and a sense of wasted time. If my time is distinct from your time, and you end up wasting my time by valuing your own, you have robbed me of my resource (time). When you value your own time instead of my time, you have effectively stolen minutes (or hours) from me. We see these attitudes in abundance.

However, if we shift perspectives and see our time as intertwined with one another's, then we are all investing our time in other people's circumstances. Communication scholar Sarah Sharma argues that the contradictions of our digital age center on the moral imperatives of "using our time wisely" while critiquing those who must negotiate with the social structures (like using food stamps) that demand that they use their time in specific ways.

Our moral perspectives around using time wisely are founded on the ways we value individual time. I must use my time wisely, and if you delay me or make me wait, you are impeding my ability to meet that social expectation of productivity. Sharma argues that if we see our time as intertwined with one another's, a new moral imperative emerges. If my time is bound with yours, it benefits me to see you use your time well or, in contrast, to help you combat the social structures that force you to spend your time in ways that put you at a disadvantage. Sharma calls for us to have a "temporal awareness" of the ways that all our time is intertwined, but often allocated unevenly to different people. If we don't foster this kind of awareness, she argues, we risk managing our own time in a way that "has the potential to further diminish the time of others."[11] Waiting can be the thing we study to see how things like racial and class inequalities force people to live time in a different way, further emphasizing their marginal positions. As I waited in line behind the woman attempting to save money and optimize her government resources, I was aware of the vastly different ways in which time becomes a strategy to create divisions between people. Our irritation at people making us wait is a distraction from the larger social structures that created this situation.

When we feel irritated or frustrated at our wait times, it may also reveal the pace of our lives at that moment. I often feel rushed. There are days whose pace seems to be at breakneck speed. When the rhythm of my day is fast, pushing me to jump from one item on my to-do list to another, and suddenly I'm confronted by a roadblock to accomplishing my goals, I get frustrated and angry. Though I live in the Washington, DC, area now, I spent most of my life in Los Angeles and started driving at age sixteen. All California drivers share the experience of being stuck behind a slow driver on the road, doing everything legally possible to get around that driver, only to be sitting next to him at the next red light. For all our stress to swerve around the driver who was slowing us down, we ended up at the exact same place. Along these lines, I've worked to recognize that my pace often does not get me where I want to go any faster, but only adds to the stress of trying to get there. I also work to recognize that if my time is collective, my attempts at getting ahead in line can be at the expense of other peoples' time.

If we work toward an awareness of time as collective rather than individual, we can come to understand wait time as an investment in the social fabric that connects us. My patience with someone like the woman at the grocery store who has to account for every dollar and pay with food stamps is an investment of my time in her situation. As we invest time in other people through waiting, we become stakeholders in their situations. This has the radical potential to build empathy and to inspire a call for social change, as we realize that not everyone is afforded the same agency for how time is used.

There are times when we should wait and see the benefits of waiting; however, there are times when waiting needs to be resisted. Waiting can be a tool of the powerful to maintain the status quo by forcing people to invest their time in ways that inhibit their ability to transform their situation. Many examples demonstrate the kinds of waiting that reinforce the power dynamics in a society. From the long-delayed recovery efforts and federal dollars following Hurricane Katrina in 2005 or the perpetually delayed recovery for Puerto Rico and other Caribbean islands after Hurricane Maria in 2017, to the long commute times between home and job (often, jobs) imposed on many people below the poverty line, unequal access to time is revealed in the different ways people are forced to wait. Many social justice advocates like Angela Davis and Michelle Alexander point to prisoners like those sitting in San Quentin as prime examples of those who are forced to wait unjustly. The "prison industrial complex," as Davis terms it, is fueled by racial inequality that targets African Americans more than any other population.[12] In this example, wait times are strategies of the powerful to maintain the status quo of power relationships in the social order.

Looking at these two tactics of waiting—seeing wait times as collective rather than individual and looking at who benefits from imposing wait times on us—can make waiting a lens through which we understand society. These tactics have the potential to connect us and allow us to enact social change. We have seen in earlier chapters other tactics that build on these foundational approaches to waiting. For example, our connection to buffering icons demonstrates how these wait times can often reveal our hopes for what might come on the other side of waiting. The buffering icon, though reviled because it makes us wait, is actually a design the presents hope and desire for a different future. The in-between moment of refreshing a social media feed or dating profile on your phone is the pause that is full of promise. The news or messages that might arrive on the other side of waiting reveal our dreams about possible futures different from our present. These desires can teach us about our hopes and also tell us about our present situation. Why do we want the future to be different, and in what ways? How can we begin not only identifying our hopes for the future but building bridges toward those hopes? These moments of waiting can be the pauses that help us reflect on what the future ought to be.

Yet the tech industry resists such pauses as much as possible. As I observed in the introduction, by reducing wait times and holding our attention longer, companies profit. Are we at a place where wait times need to be intentionally built into our systems to help us have these moments of pause and reflection? Would we be better people if our technologies helped us pause more often to reflect on our current situations and how we can begin creating the futures that we long for?

Some believe that this approach is necessary. Marina Abramović's artwork has often confronted the impact of the digital age on our attention, and with a 2015 piece she framed wait time as an antidote for our attention-weary, frantic lives. A collaboration with pianist Igor Levit, Abramović's piece *Goldberg* is a performance of Bach's *Goldberg Variations* that is preceded by a mandatory thirty-minute wait. Attendees arriving at the Park Avenue Armory in New York City exchanged their tickets for locker keys. After finding their lockers, they had to place all their belongings in the locker, including mobile devices and watches. Each audience member was then given noise-canceling headphones before entering the performance space. The Armory is a massive venue, sometimes described as a beautiful, cavernous aircraft hangar. The audience sat in white cloth deck chairs, as might be found on a beach or on a cruise ship. A gong rang as a call to put on the headphones, and the performance began with all attendees sitting in complete silence for thirty minutes. At the end of this time, Levit began his performance of the Goldberg Variations. For Abramović, imposing this wait time on the audience was a way of creating a discipline of attentiveness to prepare them for the piece. "The moment you don't have your phone and you don't have your watch to check if you're sitting there for five minutes or ten, it just gives you a completely different state of mind."[13] Zachary Woolfe, reviewing the piece for the *New York Times,* described dozing briefly, and experiencing "a few moments of deep, uncanny calm." Each person leaned back in a chair, encouraging an upward gaze to the "hall's vast arched ceiling. Bach was lent some of the childlike wonder of a planetarium star show." Woolfe continued: "An evening that could have—should have—felt ridiculous was instead . . . strangely enchanting."[14]

Abramović developed her piece in response to the constant distractions of the digital age. We are so compelled by the "attention merchants" that Tim Wu described that we require a bit of help to carve out these times of silence and contemplation. Woolfe's reflection on the piece kept returning him to the "luxuriously quiet" half-hour wait that preceded Levit's piano performance. Jen Poyant, producer of the radio show *Note to Self,* described how riveted she was in the spectacle of an entire audience dressed to the nines, wearing noise-canceling headphones: "There's no phones, nobody's moving. We were all collectively doing this one action, which is non-action, together."[15] She found herself attentive to her own heartbeat and breath. By the end of the thirty minutes, the entire audience seemed settled. In reflecting on this wait time in his review of the performance, Woolfe realized that he had never considered "silence as a commodity, one far more accessible to wealthy New Yorkers than to poor ones."

The piece may have been inspired in part by the work of John Cage, a composer whose piece 4'33" is a musical composition made up of four minutes and thirty-three seconds of silence (or, more accurately, the sounds of the space, of the bodies of the audience members, unique to each person and different each time it's performed). Both Cage's *4'33"* and Abramović's/Levit's *Goldberg* fit with the argument that I have presented throughout this book: Silence is content. It is not a prelude to content, but content itself. These silences are commentary on the contemporary desire to flee from moments of pause and waiting. They are commentary on our lack of silence and fear of the gaps that we have to fill with meaning. By embracing these pauses and imposed wait times, we are given the chance to reflect on the present rather than orient ourselves to some future object of desire.

Whether or not the digital age will require wait times built into our systems in ways that mirror Abramović and Levit's *Goldberg*, we must confront the effects of our disdain toward waiting. We must reckon with the loss of wait times and the ways this loss has transformed our own creative capacities and connections with the present moment. If we can build tactics for waiting—tactics that recognize how our time is deeply intertwined with the time of others—we can become advocates for the value of waiting. When we recognize the value of wait times, but also the potential ways in which wait times can disempower, we become students of waiting. We unpack waiting's power and promise, its value and its danger. Waiting pulls us into the present unlike any other experience of time. In waiting, we realize that this moment is meaningful as it exists, not as some step toward a future moment. Waiting is present tense, and its meanings are full of the potential to transform the ways we see the world. Each moment is its own experience and its own fulfillment.

NOTES

INTRODUCTION

1. Scholars of mobile phone culture have traced these ringing practices to a range of places around the world, especially countries where mobile costs were prepaid or users were charged per text or per call. Sometimes, a ring was a signal for a person to call back on a landline phone. Other times, a ring was a signal for a preestablished action, coordinated beforehand such as, "When I ring you, I'm downstairs to pick you up." This practice is called "beeping" in many countries. Naomi Baron traces this practice back to the era when AT&T had a monopoly on telephones in American households. During this time, when prices were incredibly high to make a phone call, people would use a single ring to communicate a message such as "if you were traveling and wanted to let the folks back home know you had arrived safely." See Naomi S. Baron, *Always On: Language in an Online and Mobile World* (Oxford: Oxford University Press, 2008), 132.

2. Renato Nicassio, "Quel Che Resta di uno Squillo" [What remains of a ring], Il Blog Struggente di un Formidabile Genio, March 2, 2015, https://ilblogstruggentediunformidabilegenio.wordpress.com/2015/03/02/quel-che-resta-di-uno-squillo/.

3. For a good primer on nonverbal communication, see Marjorie Fink Vargas, *Louder than Words: An Introduction to Nonverbal Communication* (Ames: Iowa State University Press, 1986).

4. Edward T. Hall, *The Silent Language* (Garden City, NY: Doubleday, 1959), 23.

5. Jessi Klein, *You'll Grow Out of It* (New York: Grand Central Publishing, 2016), 157.

6. Jessi Klein, "The Dress," The Moth, April 17, 2014, https://themoth.org/stories/the-dress.

7. Aziz Ansari with Eric Klinenberg, *Modern Romance* (New York: Penguin, 2015), 3.

8. LadyLayla, "Ghoster," Urban Dictionary, July 5, 2015, https://www.urbandictionary.com/define.php?term=ghoster&defid=8333139.

9. Alex Mayyasi, "Which Generation Is Most Distracted by Their Phones?" Pricenomics, February 26, 2016, https://priceonomics.com/which-generation-is-most-distracted-by-their/.

10. Patrick Nelson, "We Touch Our Phones 2,617 Times a Day, Says Study," Network World, July 7, 2016, https://www.networkworld.com/article/3092446/smartphones/we-touch-our-phones-2617-times-a-day-says-study.html.

11. Quoted in Blake Snow, "What Would a World Without Internet Look Like?" *Atlantic*, April 5, 2016, https://www.theatlantic.com/technology/archive/2016/04/a-world-without-internet/476907/.

12. This began on a global scale in 2009, though the United States started texting more and talking less two years earlier in 2007. See Ansari with Klinenberg, *Modern Romance*, 38.

13. See David Henkin, *The Postal Age: The Emergence of Modern Communications in Nineteenth-Century America* (Chicago: University of Chicago Press, 2006), 18–24.

14. Historian Cameron Blevins's work focuses on the geographies of these early post offices, as seen in "The Postal West: Spatial Integration and the American West, 1865–1902," Ph.D. diss., Stanford University, 2015, and his project in collaboration with Jason Heppler visualizing these post offices, Geography of the Post: U.S. Post Offices in the 19th Century West, http://cameronblevins.org/gotp/.

15. Allan Pred, *Urban Growth and the Circulation of Information: The United States System of Cities, 1790–1840* (Cambridge: Harvard University Press, 1973), 13.

16. This correspondence is archived at the Virginia Military Institute's archives in the Richard H. Adams, Jr., papers, 1862–1866, which can be accessed at https://archivesspace.vmi.edu/repositories/3/resources/438.

17. The gold rush began the year after five- and ten-cent stamps were introduced. See Henkin, *The Postal Age*, 119–147.

18. Harold Schweizer, *On Waiting* (New York: Routledge, 2008), 16–17.

19. Todd Hoff, "Latency Is Everywhere and It Costs You Sales—How to Crush It," High Scalability, July 25, 2009, http://highscalability.com/latency-everywhere-and-it-costs-you-sales-how-crush-it.

20. Greg Linden, "Marissa Mayer at Web 2.0," Geeking with Greg, November 9, 2006, http://glinden.blogspot.com/2006/11/marissa-mayer-at-web-20.html.

21. Schweizer, *On Waiting*, 16.

22. Philosopher Martin Heidegger calls this the shift from ready-to-hand to present-at-hand. See *Being and Time* (Oxford: Wiley-Blackwell, 1962), 99.

23. Henri Bergson, *Duration and Simultaneity: With Reference to Einstein's Theory* (New York: Bobbs-Merrill, 1965), 42.

24. Michael G. Flaherty, *The Textures of Time: Agency and Temporal Experience* (Philadelphia: Temple University Press, 2011), 14–35.

25. Sarah Sharma, *In the Meantime: Temporality and Cultural Politics* (Durham, NC: Duke University Press, 2014).

26. Mark C. Taylor, *Speed Limits: Where Time Went and Why We Have So Little* (New Haven: Yale University Press, 2014), 5–6.

CHAPTER 1. WAITING FOR WORD

1. Howard Rheingold, *Smart Mobs: The Next Social Revolution* (New York: Basic, 2002), xi.

2. Aaron Herald Skabelund, *Empire of Dogs: Canines, Japan, and the Making of the Modern Imperial World* (Ithaca, NY: Cornell University Press, 2011), 89.

3. See Rich Ling, *New Tech, New Ties: How Mobile Communication Is Reshaping Social Cohesion* (Cambridge: MIT Press, 2008), 3.

4. Marshall McLuhan, *Understanding Media: The Extensions of Man* (New York: Routledge, 1964), 22–35.

5. Kent C. Berridge and Terry E. Robinson, "What Is the Role of Dopamine in Reward: Hedonic Impact, Reward Learning, or Incentive Salience?" *Brain Research Reviews* 28 (1998): 309–369.

6. Derrida spells it différance, which is a difference that can be spotted only in the text. He uses this orthography as a way of showing the power of a text over oral communication. Jacques Derrida, *Of Grammatology*, trans. Gayatri Chakravorty Spivak (Baltimore: Johns Hopkins University Press, 1998), 158–159.

7. Hitoshi Takayanagi et al., "How Stranded Commuters in Tokyo Returned Home After the Great East Japan Earthquake," *Journal of JSCE* 1 (2013): 470.

8. The term *garakei* is a combination of the words Galapagos and the Japanese word for phone, *keitai*. The "Galapagos" portion of the term derives from the notion that Japan's technological advancements thrive in isolation, with little direct exchange between the island nation and other countries. Just as the finches Darwin studied on the Galapagos Islands, which evolved divergently into new species without reproduction across islands, Japan's technologies were free to evolve in unique ways, far beyond the pace in other technologically advanced countries like the United States.

9. See Claudia Hammond, *Time Warped: Unlocking the Mysteries of Time Perception* (New York: Harper Perennial, 2013), 18–22.

10. Richard A. Muller, *Now: The Physics of Time* (New York: Norton, 2016).

11. Katherine Wells, "Where Time Comes From," *Atlantic*, February 27, 2014, https://www.theatlantic.com/video/index/358609/where-time-comes-from/.

12. Alexis McCrossen, *Marking Modern Times: A History of Clocks, Watches, and Other Timekeepers in American Life* (Chicago: University of Chicago Press, 2013), 15.

13. Approaching this question, I've utilized some of the methods drawn from the field of media archaeology, which was established by scholars like Friedrich Kittler, Erkki Huhtamo, Lisa Gitelman, Siegfried Zielinski, and Wolfgang Ernst. See Friedrich Kittler, *Gramaphone, Film, Typewriter*, trans. Geoffrey Winthrop-Young and Michael Wutz (Stanford: Stanford University Press, 1999); Erkki Huhtamo, *Illusions in Motion: Media Archaeology of the Moving Panorama* (Cambridge: MIT Press, 2013); Lisa Gitelman, *Always Already New: Media, History, and the Data of Culture* (Cambridge: MIT Press, 2006); Siegfried Zielinski, *Deep Time of the Media: Toward and Archaeology of Hearing and Seeing by Technical Means*, trans. Gloria Custance (Cambridge: MIT Press, 2006); Wolfgang Ernst, *Digital Memory and the Archive*, ed. Jussi Parikka (Minneapolis: University of Minnesota Press, 2013).

CHAPTER 2. INSTANT MESSAGES AND PNEUMATIC TUBES

1. E. B. White, *Here Is New York* (New York: Harper, 1949), 31.

2. Esther Milne, *Letters, Postcards, Email* (New York: Routledge, 2010), 12–13. See also N. Katherine Hayles, "The Condition of Virtuality," in *Language Machines: Technologies of Literary and Cultural Production*, ed. Jeffrey Masten, Peter Stallybrass, and Nancy Vickers (New York: Routledge, 1997), 183–206.

3. Marshall McLuhan, *Understanding Media: The Extensions of Man* (New York: Routledge, 1964), 3.

4. Victor Sayo Turner, "The Strain of Always Being on Call," *Scientific American*, January 5, 2016, https://www.scientificamerican.com/article/the-strain-of-always-being-on-call/.

5. Douglas Rushkoff, *Present Shock: When Everything Happens Now* (New York: Current, 2013), 74.

6. Molly Wright Steenson, "Interfacing with the Subterranean," *Cabinet* 41 (2011): 86.

7. See Steve Redhead, *Paul Virilio: Theorist for an Accelerated Culture* (Toronto: University of Toronto Press, 2004), 50.

8. Tom Standage, *The Victorian Internet: The Remarkable Story of the Telegraph and the Nineteenth Century's On-Line Pioneers* (New York: Bloomsbury, 1998), 2.

9. Ibid., 3.

10. Holly Kruse, "Pipeline as Network: Pneumatic Systems and the Social Order," in *The Long History of New Media: Technology, Historiography, and Contextualizing Newness*, ed. David W. Park, Nicholas W. Jankowski, and Steve Jones (New York, Peter Lang: 2011), 218.

11. See Jay Lampert, *Simultaneity and Delay: A Dialectical Theory of Staggered Time* (New York: Bloomsbury, 2012).

12. See Shannon Mattern, "Puffs of Air: Communicating by Vacuum," in *Air: Alphabet City Magazine*, ed. John Knechtel (Cambridge: MIT Press, 2010): 42–57.

13. Andrew Blum, *Tubes: A Journey to the Center of the Internet* (New York: Ecco, 2012), 174.

14. See Shannon Mattern, *Code and Clay, Data and Dust: Five Thousand Years of Urban Media* (Minneapolis: University of Minnesota Press, 2017).

CHAPTER 3. SPINNING IN PLACE

1. For an excellent analysis of the rhythms of urban life, see Henri Lefebvre, *Rhythmanalysis: Space, Time and Everyday Life*, trans. Stuart Elden and Gerald Moore (New York: Continuum, 2004).

2. Neta Alexander, "Rage Against the Machine: Buffering, Noise, and Perpetual Anxiety in the Age of Connected Viewing," *Cinema Journal* 56, no. 2 (2017): 1.

3. Fiona Fui-Hoon Nah, "A Study on Tolerable Waiting Time: How Long Are Web Users Willing to Wait?" *Behavior and Information Technology* 23, no. 3 (2004): 153–163.

4. See Brad Myers, "The Importance of Percent-Done Progress Indicators for Computer-Human Interfaces," *Proceedings of CHI* (ACM SIGCHI, 1985): 11–17.

5. See Woojoo Kim, Shuping Xiong, and Zhuoqian Liang's discussion of "power" and "inverse power" icons in "Effect of Loading Symbols of Online Video on Perception of Waiting Time," *International Journal of Human-Computer Interaction* DOI: 10.1080/10447318.2017.1305051.

6. Alex Stone, "Why Waiting Is Torture," *New York Times*, August 18, 2012, http://www.nytimes.com/2012/08/19/opinion/sunday/why-waiting-in-line-is-torture.html.

7. Ben Shneiderman, "Response Time and Display Rate in Human Performance with Computers," *Computing Surveys* 16, no. 3 (1984): 265–285.

8. See also Donald Norman, *The Design of Everyday Things* (New York: Basic, 1988), 9–13; James J. Gibson, *The Ecological Approach to Visual Perception* (New York: Psychology Press, 2015), 119–135.

9. Allucquére Rosanne Stone, "Split Subjects, Not Atoms; or, How I Fell in Love with My Prosthesis," *Configurations* 2, no. 1 (1994): 178.

10. Anthony Giddens, *The Consequences of Modernity* (Stanford: Stanford University Press, 1990), 27–36.

11. Alex Stone, "Why Waiting Is Torture."

12. Russell L. Ackoff and Daniel Greenberg, *Turning Learning Right Side Up: Putting Education Back on Track* (Saddle River, NJ: Prentice Hall, 2008), 41.

13. Benjamine H. Snyder, "From Vigilance to Busyness: A Neo-Weberian Approach to Clock Time," *Sociological Theory* 31, no. 3 (2013): 243–266.

14. Claudia Hammond, *Time Warped: Unlocking the Mysteries of Time Perception* (New York: Harper Perennial, 2013), 280–288.

15. Dan Morse, "Montgomery Man Gets 9 Years in Stabbing at Post Office," *Washington Post*, May 13, 2013, https://www.washingtonpost.com/local/montgomery-man-gets-nine-years-for-stabbing-at-post-office/2013/05/13/c535110e-bbf5-11e2-97d4-a479289a31f9_story.html.

16. Pierre Bourdieu, *Pascalian Meditations* (Stanford: Stanford University Press, 1997), 228.

17. Javier Auyero, "Patients of the State: An Ethnographic Account of Poor People's Waiting," *Latin American Research Review* 46, no. 1 (2011): 11.

18. Ibid., 24.

19. Bourdieu, *Pascalian Meditations*, 228.

20. Jimena Canales argues in *A Tenth of a Second: A History* that this measurement of time was a key force in shaping modern society. Technologies that were able to detect and measure a tenth of a second, coupled with new scientific discoveries of the time required for a sensation to move from nerve endings to the brain (roughly a tenth of a second), shaped how we understood bodies and the knowledge created in this new world. René Descartes' argument "I think therefore I am" came under significant scrutiny once the time lag between stimulus and response was identified. This discovery shifted entire fields of thought, belief systems, and scientific approaches. See Jimena Canales, *A Tenth of a Second: A History* (Chicago: University of Chicago Press, 2009).

21. Kim, Xiong, and Liang, "Effect of Loading Symbols," 7.

22. Mark Wilson, "The UX Secret that Will Ruin Apps for You," *Fast Company*, July 6, 2016, https://www.fastcodesign.com/3061519/the-ux-secret-that-will-ruin-apps-for-you.

23. Roland Barthes, *A Lover's Discourse: Fragments*, trans. Richard Howard (New York: Hill and Wang, 1978), 38–40.

24. Ibid., 40.

25. Colin Campbell, *The Romantic Ethic and the Spirit of Modern Consumerism* (Oxford: Blackwell, 1987), 77–95.

CHAPTER 4. SPACE SIGNALS

1. Nicholas Carr, *The Shallows: What the Internet Is Doing to Our Brains* (New York: Norton, 2010), 6.

2. Ibid., 141–142.

3. Walter J. Ong, *Orality and Literacy: The Technologizing of the Word* (New York: Routledge, 1982, 2002), 35.

4. See Firas Khatib et al., "Crystal Structure of a Monomeric Retroviral Protease Solved by Protein Folding Game Players," *Nature Structural and Molecular Biology* 18 (2011): 1175–1177.

5. Michael J. Neufeld, "First Mission to Pluto: The Difficult Birth of New Horizons," Smithsonian National Air and Space Museum, July 10, 2015, https://airandspace.si.edu/stories/editorial/first-mission-pluto-difficult-birth-new-horizons.

6. Radio waves travel at the speed of light when in a vacuum, but more slowly in other environments.

7. Robert J. Waldinger et al., "Security of Attachment to Spouses Late in Life: Concurrent and Prospective Links with Cognitive and Emotional Wellbeing," *Clinical Psychology Science* 3, no. 4 (2015): 516–529.

8. Clayton Anderson, *The Ordinary Spaceman: From Boyhood Dreams to Astronaut* (Lincoln, NE: University of Nebraska Press, 2008), 23.

9. Chet Van Duzer, *Sea Monsters on Medieval and Renaissance Maps* (London: British Library, 2013).

10. Lisa Messeri, "We Need to Stop Talking About Space as a 'Frontier,'" *Slate*, March 15, 2017, http://www.slate.com/articles/technology/future_tense/2017/03/why_we_need_to_stop_talking_about_space_as_a_frontier.html.

11. Quoted in Mark Frauenfelder, "Imagining the Future Is Creating the Future: What I Learned After Spending Two Days with Futurists and Positive Psychologists," *Institute for the Future*, May 26, 2016, https://medium.com/institute-for-the-future/imagining-the-future-is-creating-the-future-81b011d16a3. See also Jane McGonigal, *Reality Is Broken: Why Games Make Us Better and How They Can Change the World* (New York: Penguin, 2011).

12. Sean O'Kane, "NASA Releases Even More of Its Fantastical Space Tourism Posters," The Verge, February 9, 2016, https://www.theverge.com/2016/2/9/10955126/nasa-space-posters-travel-tourism-planets.

13. Andrew Good, "Lasers Could Give Space Research Its 'Broadband' Moment," *JPL News*, February 14, 2017, https://www.jpl.nasa.gov/news/news.php?feature=6746.

14. See Marisa Cohn, "'Lifetime Issues': Temporal Relations of Design and Maintenance," *Continent* 6, no. 1 (2017): 4–12.

15. Lincoln J. Wood, "The Evolution of Deep Space Navigation: 1962–1989," *Proceedings of the 31st Annual AAS Guidance and Control Conference*, February 1–6, 2008, Breckenridge, Colorado, 6.

16. Douglas J. Mudgway, *Uplink-Downlink: A History of the Deep Space Network, 1957–1997* (Washington, DC: NASA, 2001), 541.

17. Quoted in Josie Glausiusz, "Living in an Imaginary World," *Scientific American*, January 1, 2014, https://www.scientificamerican.com/article/living-in-an-imaginary-world/.

18. Manoush Zomorodi, *Bored and Brilliant: How Spacing Out Can Unlock Your Most Productive and Creative Self* (New York: St. Martin's, 2017), 6.

19. James L. McGaugh, "Time-Dependent Processes in Memory Storage," in *Science* 153, no. 3742 (1966): 1351.

20. Ibid., 1357.

21. David A. Sousa, *How the Brain Learns* (Thousand Oaks, CA: Corwin, 2016), 49–52.

22. Ibid., 57.

23. Nikolay Vadimovich Kukushkin and Thomas James Carew, "Memory Takes Time," *Neuron* 95 (2017): 264.

CHAPTER 5. A DELAYED CROSSING

1. See Paul Ceruzzi, *Internet Alley: High Technology in Tysons Corner, 1945–2005* (Cambridge: MIT Press, 2011), 1–17.

2. Ed Malles, ed., *Bridge Building in Wartime: Colonel Wesley Brainerd's Memoir of the 50th New York Volunteer Engineers* (Knoxville: University of Tennessee Press, 1997), 89.

3. Ibid., 89.

4. Abigail Van Buren, "Wartime Correspondence Is Treasure Trove of History," Dear Abby, November 11, 1998, http://www.uexpress.com/dearabby/1998/11/11/wartime-correspondence-is-treasure-trove-of.

5. David Henkin, *The Postal Age: The Emergence of Modern Communications in Nineteenth-Century America* (Chicago: University of Chicago Press, 2006), 138.

6. Michael Todd, "A Short History of Home Mail Delivery," *Pacific Standard*, February 6, 2013, https://psmag.com/economics/a-short-history-of-mail-delivery-52444.

7. James M. McPherson, *Battle Cry of Freedom: The Civil War Era* (Oxford: Oxford University Press, 1988), 38.

8. Malles, *Bridge Building in Wartime*, 91–92.

9. Ibid., 93; emphasis in original.

10. Francis Augustín O'Reilly, *The Fredericksburg Campaign: Winter War on the Rappahannock* (Baton Rouge: Louisiana State University Press, 2003), 66.

11. Malles, *Bridge Building in Wartime*, 112.

12. Chris Mackowski and Kristopher D. White, "Before the Slaughter," *Hallowed Ground Magazine*, Winter 2012, http://www.civilwar.org/hallowed-ground-magazine/winter-2012/before-the-slaughter.html.

13. Daniel E. Sutherland, *Fredericksburg and Chancellorsville: The Dare Mark Campaign* (Lincoln: University of Nebraska Press, 1998), 35–36.

14. O. B. Curtis, *History of the Twenty-Fourth Michigan of the Iron Brigade* (Detroit: Winn and Hammond, 1891), 125.

CHAPTER 6. MARKS OF UNCERTAINTY

1. M. T. Clanchy, *From Memory to Written Record: England, 1066–1307* (Oxford: Wiley-Blackwell, 2013), 311.

2. Ibid., 295.

3. Walter Ong, *Orality and Literacy: The Technologizing of the Word* (New York: Routledge, 1982), 94.

4. Details can be read in Matthew Greenhall, "'Three of the Horsemen': The Commercial Consequences of Plague, Fire, and War on British East Coast Trade, 1660–1674," *International Journal of Maritime History* 24, no. 2 (2012): 97–126.

5. Quoted in Clanchy, *From Memory to Written Record*, 53.

6. Ibid.

7. Translated from the Latin inscription "Ortu magna, viro major, sed maxima partu, Hic iacet Henrici filia, sponsa, parens." See Elizabeth Danbury, "Queens and Powerful Women: Image and Authority," in *Good Impressions: Image and Authority in Medieval Seals*, ed. Noël Adams, John Cherry, and James Robinson (London: British Museum Press, 2008), 23.

8. Ibid.

9. Brigitte Miriam Bedos-Rezak, "In Search of a Semiotic Paradigm: The Matter of Sealing in Medieval Thought and Praxis (1050–1400)," in Adams, Cherry, and Robinson, *Good Impressions*, 2.

10. Walter Benjamin, "The Work of Art in the Age of Mechanical Reproduction," *Illuminations: Essays and Reflections*, ed. Hannah Arendt, trans. Harry Zohn (New York: Harcourt Brace, 1968), 217–251.

11. See Bedos-Rezak, "In Search of a Semiotic Paradigm," 3.

12. Ibid.

13. Clanchy, *From Memory to Written Record*, 80.

14. Marshal McLuhan, *The Gutenberg Galaxy: The Making of Typographic Man* (Toronto: University of Toronto Press, 1962), 18.

15. For a similar analysis on the censorship of maps, see Trevor Paglan, *Blank Spots on the Map: The Dark Geography of the Pentagon's Secret World* (New York: Penguin, 2009).

16. For an excellent study on such tactics, see Rita Raley, *Tactical Media* (Minneapolis: University of Minnesota Press, 2009).

17. Victor Mair, "Signature vs. Seal," *Language Log,* October 1, 2012, http://languagelog.ldc.upenn.edu/nll/?p=4230.

18. Akemi Nakamura, "'Hanko' Fate Sealed by Test of Time," *Japan Times,* August 21, 2007, https://www.japantimes.co.jp/news/2007/08/21/reference/hanko-fate-sealed-by-test-of-time/#.WcVnqdOGPMU.

CHAPTER 7. FIRST MESSAGES

1. Sherry Turkle, *Reclaiming Conversation: The Power of Talk in a Digital Age* (New York: Penguin, 2015), 3.

2. N. G. Bultin, *Economics and the Dreamtime: A Hypothetical History* (Cambridge: Cambridge University Press, 1993), 9.

3. Dave Johnston, phone interview, May 6, 2016.

4. See Jason Farman, *Mobile Interface Theory: Embodied Space and Locative Media* (New York: Routledge, 2012), 1.

5. Cornelia Roemer, "The Papyrus Roll in Egypt, Greece, and Rome," in *A Companion to the History of the Book,* ed. Simon Eliot and Jonathan Rose (Oxford: Wiley-Blackwell, 2009), 84.

6. Karen Grigsby Bates, "When It Comes to Terms Like 'Colored People's Time,' Context Matters," NPR, April 13, 2016, http://www.npr.org/sections/codeswitch/2016/04/13/474069083/when-it-comes-to-terms-like-colored-peoples-time-context-matters.

7. R. H. Mathews, "Message-Sticks Used by the Aborigines of Australia," *American Anthropologist* 10 (1897): 290.

8. Giordano Nanni and Andrea James, *Coranderrk: We Will Show the Country* (Canberra: Aboriginal Studies Press, 2013), 85.

9. Lorena Allam, "Last Refuge: Remembering Coranderrk Aboriginal Station," ABC Radio National, May 11, 2008, http://www.abc.net.au/radionational/programs/hindsight/last-refuge-remembering-coranderrk-aboriginal/3261602.

10. Diane E. Barwick, *Rebellion at Coranderrk* (Canberra: Aboriginal History, 1998), 1.

11. Nanni and James, *Coranderrk,* 20.

12. See Sarah Maddison, "Indigenous Peoples and Colonial Borders: Sovereignty, Nationhood, Identity, and Activism," in *Border Politics: Social Movements, Collective Identities, and Globalization,* ed. Nancy A. Naples and Jennifer Bickham Mendez (New York: New York University Press, 2015), 153–176.

13. Simon Schaffer, et al., *The Brokered World: Go-Betweens and Global Intelligence, 1770–1820*, eds. Simon Schaffer, Lissa Roberts, Kapil Raj, and James Delbourgo (Sagamore Beach, MA: Watson, 2009), xiv.

14. Colin Hunter, Jr., "Wurundjeri Welcome to Country," YouTube Video, March 20, 2013, https://www.youtube.com/watch?v=CzLbdw3d7UU.

15. Walter Benjamin, "Theses on the Philosophy of History," in *Illuminations: Essays and Reflections*, ed. Hannah Arendt, trans. Harry Zohn (New York: Harcourt Brace, 1968), 257–258.

16. I draw this term from Lee Vinsel. See Andrew Russell and Lee Vinsel, "Hail the Maintainers," *Aeon*, April 7, 2016, https://aeon.co/essays/innovation-is-overvalued-maintenance-often-matters-more.

CHAPTER 8. TACTICS FOR WAITING

1. *The Impossible Itself*, directed by Jacob Adams (2010; Jacob Adams Productions), online video, https://search.alexanderstreet.com/preview/work/bibliographic_entity%7Cvideo_work%7C3233739.

2. Nick Mount, "On Samuel Beckett's Waiting for Godot," presentation, Innis Town Hall, University of Toronto, January 29, 2009.

3. *The Impossible Itself*.

4. Martin Esslin, *The Theatre of the Absurd*, 3rd ed. (New York: Penguin, 1983), 20.

5. Mark Weiser, "The Computer for the 21st Century," *Scientific American*, 265, no. 3 (September 1991): 94; Mark Weiser, "Creating the Invisible Interface," invited talk, *Proceedings of the 7th Annual ACM Conference on User Interface Software and Technology*, Marina del Ray, CA, November 2–4, 1994, 1.

6. See Matthew Chalmers and Areti Galani, "Seamful Interweaving: Heterogeneity in the Theory and Design of Interactive Systems," *Proceedings of the Symposium on Designing Interactive Systems*, Cambridge, MA, August 1–4, 2004; Paul Dourish and Genevieve Bell, *Divining a Digital Future: Mess and Mythology in Ubiquitous Computing* (Cambridge: MIT Press, 2014); Genevieve Bell, "The Future Is Already Here: Making Sense of 'The Digital Transformation,'" presentation, Australian Information Industry Association, Canberra, April 5, 2017.

7. This distinction between strategies and tactics is discussed by Michel de Certeau in the introduction to *The Practice of Everyday Life* (Los Angeles: University of California Press, 1984), xix–xx.

8. Quoted in Martin Greenberger, *Computers, Communications, and Public Interest* (Baltimore: Johns Hopkins University Press, 1971), 40–41.

9. Tim Wu, *The Attention Merchants: The Epic Scramble to Get Inside Our Heads* (New York: Vintage, 2016), 7.

10. See Barry Schwartz, *Queuing and Waiting: Studies in the Social Organization of Access and Delay* (Chicago: University of Chicago Press, 1975).

11. Sarah Sharma, *In the Meantime: Temporality and Cultural Politics* (Durham, NC: Duke University Press, 2014), 149.

12. Angela Y. Davis, *Are Prisons Obsolete?* (New York: Seven Stories, 2003).

13. "Marina Abramović's Method Blew Our Minds," *Note to Self,* WYNC Studios, December 9, 2015, www.wnyc.org/story/marina-abramovic/.

14. Zachary Woolfe, "Review: In 'Goldberg,' Marina Abramovic and Igor Levit Blend Classical Music and Performance Art," *New York Times,* December 8, 2015, https://www.nytimes.com/2015/12/09/arts/music/review-in-goldberg-marina-abramovic-and-igor-levit-blend-classical-music-and-performance-art.html?_r=0.

15. "Marina Abramović's Method Blew Our Minds."

ACKNOWLEDGMENTS

The true joy of writing this book was that it allowed me to work with so many talented and fascinating people around the world. My approach required that I interview sources and experts in a wide range of fields far outside my own areas of expertise. I often felt like an academic tourist or journalist exploring new fields of thought and asking many people to teach me about their own topics of study. Without their generosity, this book would have never been possible.

I'm deeply indebted to Matthew Kirschenbaum, Yonnie Kim, Andrew Blum, Larissa Hjorth, Ayaka Ito, Miho Kamezaki, Amagasa Kunikazu, Fumitoshi Kato, Keita Matsushita, Michiko Setsu, Daisuke Harashima, Kyle Cleveland, Ian Condry, James Katz, Raff Vigliante, Nancy Pope, Thomas LaMarre, Randolph Stark, Bill Creech, Michelle Tessler, Brad Myers, Hal Weaver, Mike Buckley, Alice Bowman, Chris Hersman, Steve Gribben, Andrew Good, Joe Lazio, Leslie Rowland, Jason Wenner, Rand Boyd, Liz Depriest, Amanda Bevan, Matt Greenhall, Paul Dryburgh, Elizabeth New, Elizabeth Danbury, Brigitte Miriam Bedos-Rezak, Evan Golub, Lynette Russell, Judith Ryan, Aunty Joy Murphy Wandin, and Dave Johnston.

Many people have looked over drafts of this project, and they have helped shape the book in important ways. Many have also helped me bring this work to people in a range of venues, from publications to conferences. I'm thankful to Rob Horning, Audra Buck-Coleman, Saverio Giovacchini, Thomas Zeller, Hidenori Tomita, Daniel Grinberg, Lisa Han, Nathan Jurgenson, Ian Bogost, Chris Schaberg, Zizi

Papacharissi, Steve Jones, Jesse Merandy, Javier Obregon, Katie King, Michael James Eddy, and Sabrina Baron.

My colleagues at the University of Maryland, especially in the Department of American Studies, the Design Cultures & Creativity Program, and the Human-Computer Interaction Lab, were always supportive and helped me work through ideas; I thank Sheri Parks, John Caughey, Mary Sies, Nancy Struna, Psyche Williams-Forson, Nancy Mirabal, Christina Hanhardt, Jan Padios, Janelle Wong, Jo Paoletti, Perla Guerrero, La Marr Bruce, Krista Caballero, Sue Dwyer, Traci Dula, Ben Shneiderman, Laurie Frederik, James Harding, and Porter Olsen. I'm also thankful to Julie Greene, Ira Berlin, Alene Moyer, Daryle Williams, Bonnie Thornton-Dill, Cara Kennedy, and Myeong Lee. During the writing process, I had two research assistants, Joseph Meyer and Rahul Narla. They helped me work through many aspects of this project.

This book was generously funded through an Alfred P. Sloan Foundation grant for the Public Understanding of Science and Technology. It was also funded by a Research and Scholarship Award from the Graduate School, University of Maryland, and a subvention grant from the College of Arts and Humanities.

An early version of chapter 3 appeared in *Real Life* magazine ("Fidget Spinners: How Buffer Icons Shape Our Sense of Time," June 28, 2017), and a portion of chapter 6 appeared in the journal *Media Fields* ("Surveillance from the Middle: On Interception, Infrastructure, and the Material Flows of Asynchronous Communication," 11 [2016]). I am grateful to these venues for their editorial support and their willingness to let me republish my work in this book.

I have received invaluable guidance and support from my editor at Yale University Press, Joseph Calamia, who helped me develop my ideas and hone my writing style. He has been an important partner in the process of creating this book, from the very early days of proposing the project through the very last edits. I'm also grateful to Dan Heaton for his thorough and thoughtful editorial input and to Eva Skewes for her help throughout the publication process.

Finally, my family has been supportive throughout the process of researching and writing this book. I took many trips abroad to interview sources or work in archives, spending weeks away from home while conducting researching for each chapter in the book. My wife, Susan, has helped make those trips possible. She also helped me edit the entirety of this book and served as a sounding board as I experimented with ideas and approaches.

INDEX

Page numbers in **bold** denote illustrations.